Eliza Elder Brightwen

Glimpses into plant-life

An easy guide to the study of botany

Eliza Elder Brightwen

Glimpses into plant-life
An easy guide to the study of botany

ISBN/EAN: 9783337125004

Printed in Europe, USA, Canada, Australia, Japan

Cover: Foto ©berggeist007 / pixelio.de

More available books at **www.hansebooks.com**

GLIMPSES INTO PLANT-LIFE

NEPENTHES RAFFLESIANA.

GLIMPSES

NTO PLANT-LIFE

AN EASY GUIDE TO THE STUDY
OF BOTANY

BY

MRS. BRIGHTWEN, F.E.S.

Author of " Wild Nature Won by Kindness," &c.

WITH ILLUSTRATIONS BY THE AUTHOR AND
THEO. CARRERAS

London

T. FISHER UNWIN

PATERNOSTER SQUARE

1897

DEAR SIR JOSEPH HOOKER, –

In conversation with you I have often been impressed by your conviction of the importance of inducing young people to observe the elementary facts of botany, and I have heard you express your admiration of the efforts made in this direction, in an earlier generation, by that distinguished botanist, Prof. Henslow. You have assured me of your sense of the value of independent observations made by young students, for themselves, in the simplest and clearest language.

By your own example the studies of beginners have often been led in this direction, but in this little book, which I have ventured to produce, and of which you have kindly accepted the dedication, I have not attempted to compete with even your least ambitious flights.

All I have endeavoured to do is to prepare the minds of young people for the study of botany by explaining in the simplest language some of the elementary phenomena of plant-life. It is an humble experiment, but made, as I believe, on lines which are novel so far as they go, and essentially practical.

If, as is too likely to be the case, I have fallen into any technical errors, your good nature must not be held responsible for my fault.

Believe me to be,

Yours very sincerely,

ELIZA BRIGHTWEN.

April, 1897.

. The greater portion of this work has appeared in serial form in " The Girl's Own Paper."

PREFACE

..

WHEN I was a young girl I can well remember how much I longed for some simple book that would help me to learn, not merely the name of a plant and what class and order it belonged to, but something about its life-history.

It seemed very wonderful that, if I put seeds into the ground, dry and dead as they looked, I might feel sure young plants would presently come up; that, if I planted an acorn, a young oak-tree would in due time be seen; but how all this came to pass I could not discover. I had access to an excellent library, but although I searched there hour after hour, and found many a learned book about plants, they might as well have been written in Sanskrit for all I could understand of their scientific pages. There seemed nothing suited to the mind of a thoughtful child, and although, since that long-ago time, endless books have been written for young readers and thinkers, delightful books, too, which meet the needs of those who desire information and instruction, I have not hitherto met with one that makes the careful study of plant-life really interesting and practicable for those young people who may not have a teacher to help them in their study.

It has been my aim in the little volume I now venture to send forth to my young friends, known and unknown, to supply this deficiency. I want to enable them to share the joy of spending hours in a garden learning to understand the structure of plants. I want to make them able, when they see a bud, or a root, or a twig, to know what the history of that object is, how it comes to have the shape it takes, how it developed into its present condition, and what its next form will be.

The possession of the simple facts which I have tried to make plain to every intelligence in the following pages will turn a country walk from a useless lounge into a lively object-lesson, delightful, from beginning to end, alike to teacher and taught. Nor will I apologise for the simple language which I have used, for my design has been, while taking advantage of all the latest discoveries of science, to use no terms and introduce no ideas which cannot be made intelligible to a thoughtful child. In the hope that even so humble an effort as this may not be without a use in enlarging and quickening a sense of that infinite harmony which runs through every part of the Creator's marvellous plan of nature, I put forth, not without a full sense of its inadequacy, this little volume. It has, at any rate, given me an excuse for endless hours of pleasure within the precincts of my own woods and garden.

I have to acknowledge great indebtedness to Mr. J. W. Odell, F.R.H.S., whose wide knowledge of botanical science has been of essential service in ensuring, as far as possible, the accuracy of my statements.

ELIZA BRIGHTWEN.

CONTENTS

. .

13

CHAPTER III.

CHAPTER IV.

CHAPTER V.

CHAPTER VI.

LIST OF ILLUSTRATIONS

" To me be Nature's volume broad display'd,
    And to peruse its all-instructing page ;
    Or, haply catching inspiration thence
    Some easy passage raptur'd to translate,
    My sole delight."

CHAPTER I

ADAPTATION

" My heart is awed within me, when I think
Of the great miracle that still goes on
In silence round me   the perpetual work
Of Thy creation, finished, yet renewed
For ever.   Written on Thy works, I read
The lesson of Thine own eternity."

BRYANT.

CHAPTER I

THE study of plants appears to me to be one of the most delightful and instructive that can be taken up by young people. It has this advantage over many other pursuits that it can be carried on almost everywhere, for, even if the student's lot is to live in a town, there are generally botanic gardens within reach, and visits paid in the country are made the more enjoyable when some special study can be carried on in the daily walks.

Then collections of dried leaves and flowers can be formed during the summer, and the arrangement and classification of these will provide pleasant winter occupation.

I fear that many young people are apt to consider botany a very dry study. They are naturally repelled by the long words and many technical terms used in describing plants.

It has long been my belief that the study of botany should be approached through the garden rather than the schoolroom, beginning with a country ramble which should be an object-lesson opening out endless paths for future study.

Our Heavenly Father has given us a beautiful world to live in, and, when our eyes have once been opened to observe what lies around us, nature becomes like an exquisite book of pictures, always revealing to us something new and wonderful as we turn over each fresh page.

It is suited to all ages; the baby child begins by gathering daisies and buttercups, while older children make wild-flower collections and perhaps work in their own little gardens watching the growth of seeds and slips.

The beauty of ferns and mosses is sure to lead to some painstaking study of those fascinating growths.

Later on the fact that all trees have flowers

comes as a surprise to the unobservant, and thus, when rightly guided, young people can hardly fail to love a pursuit that promises such endless sources of interest.

In the chapters that will follow on the subject of plant life, I do not purpose to write for quite young children, as my hope is that older readers will explain what is written, and make it interesting to the little ones as they walk in gardens and fields, giving as it were object-lessons on buds, leaves, and flowers, and training young minds to search for themselves into the wonders that lie around them.

How much there is to learn about, even in the simplest things, some of the succeeding chapters will endeavour to show, for example:

How young plants grow out of seeds;

How those seeds are dispersed;

How much is folded up in a bud;

How flowers are formed;

How the bark splits off different trees.

Any one of these subjects would need very careful, patient observation truly to understand it.

I stand as it were only on the threshold of
scientific research, and look with wonder at the
work of such a student as Darwin, who gave
twenty long years to observation of the common
earth-worm before he wrote his deeply interesting
book upon it.   Again, we see Sir John Lubbock
giving years of his life to the growing of seeds
and their seed leaves, in order to learn exactly
how plants begin their life, and two very thick
volumes are required to contain the vast amount
of information he has thus obtained.

These two examples will suffice to show that
the minutest objects in nature are worthy of
reverent attention, and if these chapters tend to
awaken young people to a perception of this
fact and act as a humble guide to new lines
of thought, I shall feel that they have not been
written in vain.

I fear it is impossible to explain the processes
nature is carrying on in the plant-world without
occasionally using scientific words, but, when I
am obliged to do so I shall try to explain their
meaning,[1] and when once we rightly understand

[1] See glossary at the end of the book.

an exact expression we soon begin to use it, because it is more convenient and often saves repeating a long sentence.

I would ask my readers to try and obtain from their gardens and fields the various objects mentioned at the close of each chapter, and compare them with the plates, learning all about them as they read the letterpress.

This will, I feel sure, add much interest to the study, for having something to collect and examine tends to lighten mental work and enables us better to understand descriptive writing.

In this introductory chapter I will simply take a general view of vegetable growth and its adaptation to the situation in which it is found.

In many respects plants require the same conditions as animals, birds, and insects; they must have air, food, moisture and light in order to attain healthy growth, and although they differ from animals in being usually stationary, their life is carried on in a very similar way. Let us take a forest tree as a type.

It is anchored in the soil by its roots which are its feeding organs; through them it draws

up various kinds of nourishment from the earth
in which it stands.

The roots by several chemical processes render
the elements they have taken up from the soil
fit for the nourishment of the tree ; they send
it up through the stem and branches into the
leaves, and these being the breathing organs
have essential work to do in receiving from the
air, and giving out again, certain gases which
contribute largely to maintain the life and vigour
of the tree. Thus it grows year by year, producing annually its flowers and seed, which is
the end and aim of all plant life.

We can trace another analogy with animal
life, in the necessity for pure sweet air, plants
growing in a vitiated or smoke-laden atmosphere
soon showing unmistakable signs of weakness.
The stunted hedges and trees on the fringe of
London always remind me of the poor, ill-grown
children of the slums.

Besides the plant life which we see around us in
the shape of trees, shrubs, and flowers, there are
lower and perhaps still more wonderful forms of
vegetable life affording endless fields of study.

Mosses, lichens, and fungi we are familiar
with everywhere in the country, but below these
again are such growths as the green stain [1] which
makes the tree trunks in moist places as brilliant
in colour as the leaves themselves. Looked at
through a lens we see the colour arises from a
growing plant of extremely simple form, little
more in fact than a succession of cells, each living
and increasing "after its kind."

Again, if we consider the process of fermentation, we find that when it is set up in a cask of
wine its action is due to the growth of a minute
vegetable that feeds upon the alcohol and sugar,
and by robbing the wine of those two elements
turns it into vinegar or acetic acid.

A somewhat similar growth causes the thick
jelly-like substance we sometimes find in our ink-
glass when it has been allowed to remain too long
without renewal; the minute germs floating in the
air have found the ink suitable to them, and thus
their mycelium [2] begins to form at the bottom of
the glass, to the great discomfort of the writer.

The yeast with which our bread is fermented is

[1] *Protococcus.*   First form of fungoid growth

another of these minute plants, and consists of
oval cells which multiply with great rapidity when
placed in a pan of flour, and kept in a warm
atmosphere.

By the careful study of these lower forms of
vegetable life, Pasteur, Koch, Frankland, and
others have discovered and classified the germs
or microbes,[1] as they are called, which give rise to
various diseases. In books upon the subject, their
different shapes are figured as they appear when
immensely magnified, so that we can see that which
will give rise to consumption, erysipelas, or cholera,
and one reads with deep wonderment of all that
science has ascertained of late years as to the
presence in the air of these seeds of disease
which are ever floating more or less around us.
But for the restraining hand of God, it appears
as if universal sickness and death would be our
fate.

Leaving these lower forms of growth, we may
consider the three divisions into which plants are
naturally classed as to their duration of life.

Annuals are those which grow and flower, and

[1] Small living atoms.

form their seeds in one year, within which their
life-history is closed.

Biennials produce leaves only in the first year ;
by their aid they lay up stores of nutriment in the
form of tuberous roots, on this food they can exist
through the winter, produce flowers the following
summer, perfect their seeds, and then die.

To this class we owe such useful plants as the
carrot, parsnip, beetroot, and many others which
afford us such nourishing vegetable diet.

Perennial plants live on for an indefinite number
of years, flowering annually, in some cases dying
down to the root in autumn, and producing fresh
foliage the following year.

Water plants seldom have a fixed root, but remain floating, borne up and kept in position by the
water, their roots being the means by which, in
conjunction with the leaves, they derive nourishment from air and water. It is well worth while
to observe the two forms of leaves in the water
buttercup. Those on the surface are three-lobed,
flat, and round, they absorb from the air such gases
as the plant requires ; while the leaves beneath
the surface are divided into threads so as to offer

no obstruction to the flow of water and enable the
plant to collect needful food from the water.   It
can vary the form of its leaves according to its
requirements, since in running streams it may
often be found with the hair-like leaves only.

WATER BUTTERCUP.

On the other hand, if its seeds are sown in moist
earth, the seedlings will grow and develop those
flat leaves only which are characteristic of land
plants.   This water buttercup, therefore, gives us a

wonderful example of adaptation to surrounding influences.

Adaptation is remarkably shown in the Vallis-

VALLISNERIA.

neria, a grass-like water-plant, found in Southern Europe;[1] it grows in freshwater lakes, rooted in the

[1] It can generally be met with at naturalists' shops where aquaria are sold.

mud, and yet its flowers need to be fertilised in the
air.　In order to effect this, the small male flowers
detach themselves from their stems, and, rising
through the water, float about upon its surface.
The female flowers are borne on a stalk, spirally
twisted, so that it can uncoil and allow the flower
to reach the top of the water whether it be deep
or shallow.　There the two kinds of flowers meet,
the seeds are formed and the stem coils up again
and brings the capsule below the surface, where
it gradually matures.

The water-lily can grow a long or short stem
as the depth of the water may require to enable
its leaves to lie flat upon the surface.　I have
gathered lily flowers in my lake with stems from
four to five feet long, where the plant happened
to be growing in deep water.

In such plants as the mare's-tail (*Hippuris
vulgaris*), we find the stem specially adapted to
a submerged life.　Growing out of mud at the
bottom of a stream the plant upholds its slender
stalks by two different methods.　Inside the epidermis (or outer skin) a strand of rather tough
tissue running through the centre gives flexible

support, whilst the rest of the space is filled up
with very large air cells, which give such buoyancy
to the stems that even if they are three feet in
length they are kept upright in the water, rising
ten or twelve inches above the surface. It is a
valuable as well as a curious plant, as it has the
property of absorbing the gases emitted by stagnant water, and tends thus to purify the air.

The same power of adaptation is to be found
in sea-weeds. Those growing on rocky shores
having short fronds covered with fructification,
while out at sea, ribbons of oar-weed may be found
many yards in length, formed, like the gulf-weed,
of tough texture to bear the friction of waves and
storms.

If we were travelling in a Mexican desert, we
should find those remarkable plants which can be
so well studied in the cactus-house at Kew Gardens.
Bearing in mind that for many months the plant
must do without a drop of rain, or in fact without
moisture of any kind, it has been necessary that
the leaf-surface should be reduced to prevent loss
of moisture by evaporation, and so spines take
the place of leaves, and the stems are encased

in a thick leathery skin, which protects the plant
from the burning heat of the sun. Very little
moisture escapes through this thick green epidermis ; therefore when rain falls the plants receive
and store up their liquid food, and live sparingly
upon it during the long periods of drought, which
last for three-quarters of the year. Some of these
cacti, as we see them at Kew, are tall, straight-
stemmed plants, others low-growing rounded
masses, little spiny cushions, almost like vegetable
hedgehogs.

In the arid prairies of Texas, advantage is taken
of the watery stores of the cactus, for when other
supplies fail, its fleshy stems are cut open, and
horses and cows greedily devour the succulent
food, which answers the purpose of drink, as well
as affording nutritious fodder.

Our British spurge-plants have green leaves,
a thin epidermis, and all the ordinary characters
of the plants of a temperate region, but by comparing them with the spurges found in Madeira,
we see how climate causes adaptation to differing
conditions. One of these spurges growing in my
greenhouse has a tall column-like stem, no leaves,

and a thick leathery skin, which would enable it
to bear a hot, dry climate. It thus mimics the
giant cacti of Mexico.

We may trace another contrast in our common

CATTLEYA WALKERIANA (*a Brazilian Orchid*).

groundsel and the large succulent groundsels of the
Cape and the Canary Isles, with their thick fleshy
leaves, the difference in form and texture being
simply an expression of the wonderful modification
due to climate.

The lovely tribe of orchids make the same
provision for long periods of drought. Many of
the species live in countries where the rainy season
lasts about six months, and is succeeded by as
many months of dryness and heat.

The air-plants we obtain from these countries
have large pseudo-bulbs, that is, the stems are
enlarged so as to be storehouses of nutriment upon
which the plant exists, and by means of which
it brings out the gorgeous flowers which make
Brazilian forests such fairylands of beauty ; every
tree-branch being laden with parasitic orchids,
their lovely blossoms lasting month after month
without the aid of rain or dew, because Nature
has provided each plant with its special store of
food, and has thus adapted it to the position it is
created to adorn.

Another of these perching-plants is *Tillandsia
Usnoides*, known in Florida as Spanish moss, and
often called "old man's beard." It hangs from
the tree-branches in tufts, like grey hair, and grows
in such profusion that it is collected and used for
stuffing cushions. This curious plant has no roots,
but simply hangs from the branches, and lives like

the orchids by absorbing water from the moist air
in the humid forests where it is found.

The absorption by the long, hanging, grey roots
of the orchids in one case, and by the finely-divided
leaves and stems in the other, are both instances
of the wonderful way in which Nature "adapts"
the parts of a plant to its requirements.

It often happens that seeds, blown hither and
thither by the wind, chance to fall upon places
which are quite unsuitable to their mode of growth ;
then we have an opportunity of seeing how their
power of adaptation enables them to triumph over
almost insuperable difficulties.

I have observed a tiny plant of groundsel
growing out of a chink in a wall where there
was scarcely any soil from which it could derive
nourishment, contriving to live on, however, and
make the best of its hard lot. Its stem, which
should have been a foot high, could only attain
about two inches, and instead of dozens of leaves
it had but four, and yet it survived and even
produced two small flowers, thus touchingly dis-
playing its power of adaptation.

Another more remarkable instance which occurs

to me was that of a seedling Scotch fir, which had
rooted itself in a lump of house-leek on the top

YOUNG SCOTCH FIR GROWING IN HOUSE-LEEK.

of a garden wall. For eight years the young tree
managed to live and grow, until it became a sym-

metrical well-branched fir-tree, almost twelve inches
high. By a supreme effort it produced a crop of
miniature cones, and soon after it died from
drought and starvation, the wonder being that it
could have lived so long upon the modicum of
food the barren wall supplied, besides having to
endure at times periods of scorching heat as well
as drought. The chief interest in this example is
centred in the fact that as soon as fruit-bearing
has been attained, then, and not till then, the little
tree died, showing how persistently under all
hindrances and difficulties a plant will endeavour
to carry out the purpose of its creation.

We have seen in these instances some striking
examples of the way in which plant-life is adapted
to its surroundings. Our examples have been
such as are easy of attainment, and such as we
can verify with our own eyes; but even more
wonderful are the adaptations hidden away in
the recesses of the plant, and as we progress in
our study these arrangements of cells and tissues
will be revealed to us. In order however to see
them, and to understand their true significance,
we must proceed step by step to study the parts

of an ordinary plant ; because it is only by first
mastering all we can of one part of a plant, and
then comparing that part with other plants, that
we can hope to gain real knowledge. Accordingly
in our next chapter we shall take the root as our
starting-point, and ascertain its functions and uses,
and the part it has to play in the economy of the
plant.

Specimens to be obtained :—Green stain on tree-
bark (*Protococcus*) ; yeast ; annual, biennial, and
perennial plants ; water buttercup leaves ; vallisneria ; water-lily stems ; mare's-tail plant ; cacti ;
spurge ; orchids ; tillandsia ; plants growing in
wall crevices.

CHAPTER II

ROOTS

"While thus through all the stages thou hast push'd
  Of treeship   first a seedling, hid in grass ;
  Then twig ; then sapling ; and, as century roll'd
  Slow after century, a giant bulk
  Of girth enormous, with moss-cushion'd root
  Upheaved above the soil."

COWPER.

ROOTS

ET us begin our study of roots by considering the way in which plants obtain their nourishment from the earth, and are kept in an upright position by means of their root-fibres. These being out of sight, we may easily not be familiar with this part of the economy of plant life, but we shall soon see what important duties the roots have to fulfil, and how much they vary in character and appearance according to the soil, the climate, and the work they are required to do. The greater number of annual plants (those which live only one year) have fibrous roots, and of these we can find examples almost everywhere. A piece of groundsel or tuft of grass will answer our purpose.

4

On pulling it out of the ground we see a bunch of whitish threads or fibres springing from the crown of the plant (which is the junction between the stem and the root), and on these slender fibres are hairs which are really the active part of the root, for it is only through these hairs that the rootlets are able to absorb the liquid from the soil, the fibres simply acting as channels to convey the watery nourishment to the stem and leaves.

Common earth consists of small particles of mineral substances such as flint, chalk, or iron, and also of such vegetable matter as decayed leaves and rotten wood.

The spaces between the particles are more or less filled with air, each mineral particle being enveloped with a film of water. However dry the soil may appear, this will always be found to be the case. It may be tested by weighing in an agate balance some dry soil on a summer's day. There is a very delicate instrument called a hygroscope, which can tell us when there is the slightest amount of moisture in the air, and a clever German writer, Von Sachs,[1] has termed this

[1] Author of " Vegetable Physiology."

film of water, which gathers round earth-particles, hygroscopic water. It has been ascertained by careful experiment that it is only on this delicate watery film that the root-hairs of plants are able to feed. As these hairs drain away the hygroscopic film it is always being renewed by the free water which comes from rain and dew. The free water of the soil is constantly passing from the surface to the subsoil, and by this action plant-food, in the form of soluble earth salts, is presented to the roots. The passage of the water is of the highest service to the roots, since the warm air follows the water through the soil, and helps to oxidise the mineral particles ; these are thus rendered soluble, and are taken up by the fine films of water, and so indirectly the roots are fed. If, however, there is no outlet for the water and the soil becomes water-logged this beneficial action is retarded, and to land-roots the water is hurtful.

We can now understand why stagnant water in the ground is so injurious to plant-life, as it prevents the needful air from coming into contact with the roots, and this is the reason why farmers are careful to remove the surplus water from their fields by thorough drainage and ploughing. Roots adapt themselves very wonderfully to their situation.

This piece of grass, which we are examining, if it grew in sandy soil, would have its root-fibres covered with a downy growth to enable them the more readily to absorb every particle of moisture in the sand. Dr. Bonar speaks of the date-palm as having this same characteristic. "These palm roots are peculiarly fitted to obtain every drop of water that the sand contains: they consist of long fleshy strings or ropes, shooting straight down into the sand, in numbers quite beyond our reckoning, and extending over a large circle."

The tendency of fibrous roots to bind sand together is taken advantage of on many of our sea-coasts, where the sand blows inland and renders acres of ground sterile and useless. There, if the *Carex arenaria* (a kind of sedge) is planted, its roots will spread far and wide, interlacing and creeping through the sandy soil, until in time the latter becomes solid and no longer drifts inland.

An allied species of grass, *Psamma arenaria* (or

marrem grass) grows abundantly at Bourne-
mouth, and wishing to ascertain how far one of its
underground stems extended, with some amount of
patience I disinterred about six or seven feet of it
in a bank on the sea-shore where it was accessible.
As it seemed to have no end, I could not ascertain
its entire length.

Another instance of root growth adapting itself
to situations occurs to me. In visiting the
Cheddar Cliffs in Somersetshire I was struck by
the beauty of a plant which grew here and there
out of the crevices of the rocks. Its tufts of vivid
green leaves looked so healthy and vigorous I
could not help wondering how it could obtain
moisture enough to produce such foliage, placed
as it was high up on the dry face of a rock.

Failing to reach its roots in any other way, I
climbed up to a spot where I could remove some
of the horizontal layers of stone. At last I lifted a
flat piece of rock just above one of those plants,
and there I saw at a glance the secret of its
vigorous growth.

The roots had spread out over the surface of the
stone for a distance of eight or nine inches in a

perfectly flat layer of fine fibrous rootlets no
thicker than a sheet of paper; these would doubt-
less suck up abundant moisture whenever the rain
beat upon the rocks, and there, pressed closely
between the two layers of stone the plant has its
water-supply stored up, and is enabled to look
fresh and green when other vegetation is suffering
from drought.

In plant-life there is a marvellous variety in
root-structure. Roots differ much, not only in
form, but in texture and duration of life, so that to
gain a true knowledge of them we must carefully
examine those of herbs, shrubs, and trees, and
observation will soon teach us the fact that there
exists a close correlation between the form and
texture of the root and the size and character of
the plant. The external shape will depend principally upon whether a tap-root is developed or no.
Such, for instance, as the carrot and the dock are
those of the true tap-root character. Of branching
roots we may find endless modifications amongst
ordinary field or garden flowers from the fibrous
roots of the little groundsel to the large fleshy
tubers of the dahlia. Between these two types

there are others of an intermediate kind, but it is
possible to recognise amongst common plants the
roots belonging to one or other of the types I
have described. For the purposes of study we
may broadly group roots into classes according to
their method of collecting and absorbing food.
Thus we find one group growing in soil and
feeding upon the soluble earth salts and moisture
of the soil. Another group will be found growing
in water, like the water-lily and pond weeds. A
third group simply hangs down in space from
some perching plant like the tropical orchid, whilst
a fourth and very small group consists of parasitic
roots, of which a very common example is the
mistletoe. We will now study each of these
groups separately.

I have already spoken of some kinds of fibrous
roots, and may add that if the root of a land plant
is immersed in water, it will after a time develop a
different kind of fibre, capable of receiving nourishment from water instead of earth. I remember
seeing an instance of this in the case of a laurel
bush which grew near a well in our garden. We
had occasion to examine the water, and found that

the laurel had thrown down its roots below the
surface, where they grew luxuriantly, finely sub-
divided, of a delicate ivory white, owing to the
absence of light, and more than a yard in length.
They had adapted themselves to the duty of
absorbing water only, but had we replanted them
in earth they would have withered, from their
unfitness to take up the hygroscopic water of

CREEPING GRASS.

which I have already spoken.　On the other hand,
if the seeds of a plant formed to live in the water,
such, for instance, as the water-lily, are sown in
ordinary soil, they adapt themselves to the new
conditions, and are able to live on the hygroscopic
water they find around the particles of earth.

　Some plants send out a horizontal stem (culm)
along the ground, with a bud and some roots

growing out of it at regular intervals. Each of these joints (or nodes) takes root and forms a separate plant. What are called strawberry runners are stems of this kind, and so are the creeping stems of *Potentilla reptans.*

POA BULBOSA.

I once found a plant of the latter growing on a low wall, and, as I imagine, because it desired to reach the ground and root itself there, it had thrown down a stem a yard and a half long with eight young plants growing upon it at intervals

ready to form so many colonies when they should
reach the ground.

One may frequently find stems of various
grasses running along the ground, and taking
root at each joint. I have one such spray in my
herbarium, with twelve young plants upon it at
regular intervals.

Some plants store up nourishment in their roots,
as may be seen in one of our common seaside
grasses (*Poa bulbosa*); this soon withers after
flowering, and becoming uprooted, its bulbs, which
are like small round cheeses strung together, may
be seen blowing about in the wind.

With such a provision as this, the parent plant
is able to bear extremes of cold and drought.

It is well for us that plants have this power of
storing up their food underground, for to it we
owe such useful tubers as the potato and Jeru-
salem artichoke.

One of our native plants, the earth-nut (*Bunium
flexuosum*), has a single round tuber which is eatable when roasted, and is often dug up by children.
Long ago, when England was liable to famines,
even this small tuber was valued as a means of

eking out the labourer's daily meal. It is worth while to examine the curious divided tubers of some of our common orchises, such as the spotted orchis (*O. maculata*), or the meadow orchis (*O. morio*). The tuber which produces the leaves and flowers withers away at the end of the summer, but it leaves behind it a second tuber in which is stored up the nourishment required to enable it to bring forth leaves and flowers in the following spring.

Tubers are in reality underground stems which have thickened into rounded balls to contain plant food.

If we examine a potato we shall see that it contains true buds in the little hollows on its surface; these are called "eyes," and each of them if sown in the ground will produce a new potato plant. If a potato is left in a damp cellar, each of these eyes will send out a stem, thus proving that the "eye" has the nature of a bud. If we cut the potato in half we shall see it is of an even substance mainly composed of starch, but if we halve an onion it will be found to consist of rings or layers of a thick fleshy nature, which proves it to be a bulb and not a tuber. The onion is like a large bud growing underground, instead of on a tree branch. We can prove how similar the onion and the bud are, by searching on a lily stem for buds or bulbils, which

LILY BULBILS.

are often produced in the axils of the leaves; if we plant such a bud it will throw out fibres and become a bulbous-rooted plant. Some of our native grasses seem to have a singular power of adapting themselves to their position. For instance, the common Timothy grass *Phleum pratense*, which usually lives by means of a fibrous root, can, if needful, produce a bulb which enables it to keep living in a very dry place, but if removed to a wet soil it returns to a fibrous root. Other grasses have been observed to alter their root-growth in the same way, adapting themselves to their surroundings.

AIR ROOTS.

These absorb the watery vapour of the air; they cannot adapt themselves to live in earth, but under certain conditions they can put forth other kinds of roots that are partially adapted for growing in soil.

I may here give some personal observations about a certain *Hoya* plant that came into my possession so long ago as 1855. This much-enduring plant lived in a hanging basket for many years, in the dry air of a sitting-room. Its leaves were sometimes shrivelled from lack of water, and it never had vigour enough to produce flowers. At last, after enduring this life for twenty years, it was placed in a stove-house where the moist heat suited its requirements. Then it flowered charmingly, and

even now is showing a further degree of enterprise
by growing a bunch of fibrous roots at the end of
one of its stems.   I imagine it intends to plant
itself into another pot standing near.   I am watching it with much curiosity, because if it does this,
the old plant will prove that it has a high degree
of intelligence, and that although it remained
quiescent for so many years, it was only from lack
of opportunity to do more than quietly endure its
privations.

In tropical countries, some plants and trees such
as *Monstera* and *Philodendron* send down slender
aerial roots called lianes, many hundred feet in
length.

In the Aroid House at Kew, I remember seeing
these lianes coming down from the roof of the
house in search of water and earthy nourishment.
It seemed like actual intelligence that directed
these roots to a tank of water twenty-five feet
distant from their starting-point above. Whilst we
are considering this subject, I may mention the
curious root action of a kind of fig-tree growing in
the tropics which is sometimes known by the name
of the "Murderer."   Its seed often falls, or is

dropped by birds, amongst the leaves in the head of a palm-tree, there it begins to grow and forms root after root, gradually descending the stem of the tree and clasping it so tightly that at last the palm is strangled and falls to the ground carrying its destroyer with it, where it roots and grows into a tree.

Parasitic Roots.

As in human society there are thievish characters who live by preying upon their neighbours, so in vegetable society we find quite a number of different plants growing at the expense of others, inserting their roots into the stems and roots of trees instead of drawing their nourishment from the ground. Careful distinction must be drawn between such plants as ivy, virginian creeper, clematis, lichens, &c., which simply grow and climb on the bark of trees, and the true parasites which are nourished by the juices of the trees and plants into which their roots penetrate.

Some plants are only partially parasitic, such as the cow-wheat (*Melampyrum*) and the yellow rattle (*Rhinanthus*). These represent a very deceitful kind of growth.   To all appearance the plants are

CLOVER DODDER.

getting an honest living, the leaves are perfectly
green and capable of performing all the duties of

leaves, and yet, if we remove a little of the soil the plant will be found to be attached to, and growing from the roots of some strong kind of grass, and is deriving its nourishment from the food collected by those grass roots.

Yellow rattle grows abundantly in undrained marshy fields, where it is easy to obtain the plant, so as to examine its mode of growth. We may then go on to a clover-field and seek for that true parasite and most troublesome enemy to the farmer, the clover-dodder (*Cuscuta trifolii*). Its seeds are frequently mixed with the clover, and when sown they germinate on the surface, but the little thread-like stem, instead of entering the ground, feels about in the air until it reaches a young clover-plant. It soon clasps its victim with its fast-growing stem; as the clover grows the dodder coils around it and is carried away from the ground.

As the wiry stem gains strength, it developes a series of suckers that eat into the clover stem and rob it of the food it has collected; it lives, flowers, and grows at the other's expense. The rate of growth of the dodder exceeds that of the clover, so

5

that the latter is both exhausted and choked by its snake-like enemy.

I once sowed a patch of flax in a garden, and not knowing that it too had a parasitic enemy, I was greatly puzzled to find quantities of pinkish threads growing out of the flax stems. These threads bore round bunches of tiny flowers. All this was very pretty and interesting, but it resulted in my patch of flax becoming a mass of interlacing threads and dying a miserable death, fairly strangled by the flax dodder. Another species of *Cuscuta epilinum*, grows on furze and also on heather, it having the twine-like stems by which dodder may readily be known.

We are all familiar with the mistletoe, its leathery leaves and its white berries.

This plant grows out of the branches of poplar, hawthorn, and apple, and very occasionally upon the oak.

In France and Belgium the custom of bordering the fields with single rows of Lombardy poplars seems to favour the growth of mistletoe, for its large green bunches form quite a feature in the landscape, and cannot fail to be observed by the traveller as he journeys in the railway train. I have been told that mistletoe is sufficiently abundant to be used in Normandy as cattle-food.

If a mistletoe-berry is gently pressed upon a young branch of an apple-tree, its own viscid juice will cause it to adhere, and before long it germinates and sends its roots into the tissues of the tree; as it grows, it fuses with them, and derives all its root nourishment from the substances in the branch. Of course the tree is weakened by this parasite, the sucking roots of which disturb the flow of the sap; woody knots are apt to form, and not unfrequently the branch is killed by the intruder which has fastened upon it.

Having touched upon the four principal kinds of roots, we will now take a single root-fibre and examine it more closely. It seems scarcely possible that such a brittle, feeble thread should be able to penetrate into the ground and make its way amongst stones and sharp-edged fragments of earth without being bruised or torn. The chief friction is borne by the growing-point, and this always has for its protection a root-cap; the section of the growing-point of root-fibre given in the plate shows the outer

skin, called the epidermis, and over that is the root-
cap shaped like a thimble formed of small cells.
As they are worn away outside and become dead
tissue, owing to friction with the soil, the cells are
constantly being renewed from within. The root
is thus enabled to grow and perform its part in
maintaining the life of the plant. The presence of
this root-cap and the absence of
leaves are the marks by which a true
root is known and distinguished from
an underground stem. With a small
lens one can see this extinguisher-like
cap protecting the extreme point of
the root, and it is well to examine a
variety of specimens, and see how they
differ slightly in size and shape.

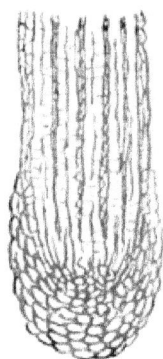

SECTION OF
ROOT CAP.

The one especial office of the root is to absorb
liquid nourishment from the soil for the benefit of
the plant, and, as I have already explained, this is
done mainly through the hairs which grow upon
the fibres of the roots. For instance, there is no
absorption in old tree-roots, such as we sometimes
see above the ground, nor in carrots, turnips, and
parsnips, but thrown out from such bulbous plants

are the fibres and their hairs which enable them to
grow to maturity. We may naturally inquire how
the solid materials in the soil, which are needful to
the growth of the stem and leaves, can possibly be
taken up by these extremely minute hairs.

We may look upon the earth as being a sort of
store-house of indigestible, unprepared plant-food
which must be altered in its character before it will
be fit for absorption by the roots. Some substances,
such as sugar, will readily dissolve in water; others,
such as starch and sand, are insoluble, but the effect
of rain-water and atmospheric air passing through
the soil, converts this insoluble dormant food into
soluble active food.

The root-hairs convey this food to the small
fibres, and through them as channels it passes on
to larger ones, until it reaches the stem and goes to
feed the growing leaves and flowers.

In order to remain in a healthy state, roots must
absorb oxygen gas, and for this reason gardeners,
when they find the soil growing caked and hard on
the surface, dig and rake the flower-borders in order
that air may freely permeate the soil and find access
to the roots of the plants.

Roots appear to be endowed with certain remarkable attributes, about which learned books have been written of late, giving the result of patient investigation as to their power of movement, the way in which they are affected by gravitation, the influence of light, and other forces.

The experiments of Darwin and other scientists have revealed very singular facts about the movements of plants. The term used to describe their motion is one we must learn, as it frequently appears in botanical works. Circumnutation we may translate as wavering around, and it well describes the curious way in which rootlets, for instance, are always moving slowly from one side to the other, describing a kind of oval zig-zag track through the earth. The fibres appear to have a discriminating power, enabling them to choose convenient crevices through which to penetrate hard soil, to avoid stone, and to seek out any attractive food which lies in their way.

As soon as roots emerge from the seed they at once turn from the light and seek to bury themselves in the earth; the plumule from which the leaves will spring has exactly the reverse tendency.

and invariably seeks the light and grows upwards. This can be proved by growing some mustard seeds on a piece of flannel about the size of a shilling, floating it on water in a saucer exposed to light from a single window; as soon as the leaves appear, they will lean towards the light, whilst the roots will point towards the dark part of the room. If a germinating seed is even placed with the root uppermost, and the plumule pointing downwards, it will very speedily right itself, the stem will turn and grow up, and the root will seek the ground.

The amazing strength of growing tree roots can be imagined when we watch a tree in full leaf during a high wind. As the terrific force of the gale sways the trunk backwards and forwards the roots are subjected to an enormous strain. Like great india-rubber cables they give and retract, and when the wind subsides we find the trunk as rigid as ever.

If my readers will seek for the specimens enumerated below, and compare them with the remarks made in this chapter, they will have such a general idea of the functions of roots as will, I

trust, enable them to enjoy the study of more
advanced works upon the subject.

Specimens to be obtained and compared with
the descriptions in this chapter :—Sedge or marrem
grass growing on a sandy sea coast ; plants growing
between layers of stone ; tree roots at the edge of
a pond ; strawberry runners ; a plant of *Potentilla
reptans* ; creeping grasses ; *Poa bulbosa* roots from
the seaside ; potato.   Earth nuts ; lily bulbils ;
Timothy grass ; cow-wheat *Melampyrum* ; yellow
rattle *Rhinanthus* ; clover dodder *Cuscuta trifolii* ;
flax dodder *Cuscuta epilinum* ; mistletoe.   Root
fibres of various plants.   Mustard seed sown on
flannel.

CHAPTER III

TREE STEMS

"If thou art worn and hard beset
 With sorrows, that thou wouldst forget,
 If thou wouldst read a lesson, that will keep
 Thy heart from fainting and thy soul from sleep,
 Go to the woods and hills!   No tears
 Dim the sweet look that Nature wears."

                                LONGFELLOW.

CHAPTER III

TREE STEMS

A WALK through a wood on a bright day in February will afford us many interesting intuitions about the growth of trees.

We are apt to think of winter as a dead season, and long for summer days once more, that we may pursue our botanical studies; but as soon as February begins there is already a secret work going on within the tree-stems, the sap is rising from the roots, and this ascent is easily to be traced if we look carefully at the trunks of those trees, such as the oak, elm, and others, which have rugged bark. The wood within is swelling; fresh layers of material will, a little later on, be added to the inner side of the bark as a result of this ascent of the sap.

75

As the bark is hard and inelastic, it cannot expand in proportion, and therefore has to crack and split in yielding to the internal pressure. If we look for these fresh cracks, we shall see the

TURKEY OAK BARK.

clean new bark within, which, before long, will harden and become of the same shade of grey as the rest of the stem.

It is at this season, too, that the plane-tree sheds off its fragments of bark in greatest quantity, as

one may plainly see in the London squares, where
this tree grows so remarkably well. Its stem is
always peeling more or less throughout the year,
and possibly that fact may be one of the reasons

SCOTCH FIR BARK.

of its flourishing so well in the midst of smoke
and fog.

Trees shed their bark in many different ways.

A reference to the illustrations will show the
concentric rings of the horse-chestnut, the square

pieces of the sycamore, which are due to the cleavage being both vertical and horizontal, the hexagonal shape of the divisions of the Scotch fir, the rugged bark of the Turkey oak, the sycamore and other species.

SNAKE-BARK MAPLE.

Where a woodpecker or a nuthatch has bored a hole into the living wood of a tree-stem, it is interesting to watch how the injury is repaired. New bark begins to form at the edges of the wound, and to this a layer is added each year, until

at last the hole is filled up, and only a scar is left
to show where it once existed.

I have been able to watch this repairing process
going on for twelve years in the case of a Turkey

TURKEY OAK STEM (*struck by lightning*).

oak, which was injured by lightning. I was watching the progress of the storm from one of our
upper windows, and happened to be looking at
this particular tree in the park, when out of a lurid
cloud above it, a streak of forked lightning

descended upon the tree, and rent off the bark of
one side from the top to the bottom, carrying away
portions of it to a distance of fifty feet or more,
leaving a white gash which looked pitiful enough

PLANE-TREE BARK.

for many months. Year by year a wave of new
bark rolls on, covering the bare place by slow
degrees, but it is never destined to be quite healed
in this case, for the inner wood was killed to some
extent by the lightning, so it has become a home

for the boring beetles, who are riddling it with holes wherein to lay their eggs.

Such a tree becomes a happy hunting ground for the woodpecker, who is attracted by the insect

HORSE-CHESTNUT BARK.

diet he finds there. The large holes he makes in getting at his prey will let in the rain, so that after a time the moist rotten wood forms a suitable place for various fungoid growths, and all these agencies work together for the destruction of the

wood until the tree becomes a hollow stem, and the
leafage above is solely produced by the sap carried
upward by the bark.

Let us inquire a little more carefully into the

WHITE POPLAR BARK.

formation of a tree-stem, and the different parts of
which it consists. Some rather hard names are
given to the four principal parts of a tree trunk
but, by reference to the plate, and by knowing the
meaning of the names, I hope they will soon be

mastered, and then our future walks in the woods
will be fuller of interest than ever, when once we
understand something about the hidden work that
is being carried on in those grand old trunks

SYCAMORE BARK.

around us. A tree may be compared to a large
manufactory. As we stand outside the building
we see the brick walls and the roof, and smoke is
coming out of the chimneys. We know that a
great deal of work is being done inside, and carts

are leaving its doors laden with the products of the
machinery within, but how the work is done we
cannot tell from the outside. We perhaps desire
to obtain this knowledge, and under the guidance
of the manager, we are taken from room to room

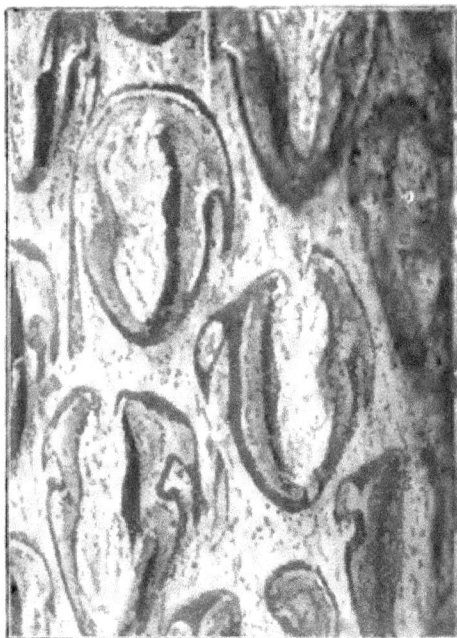

TREE-FERN BARK.

and see the marvellous processes by which raw
material is converted into exquisite fabrics, or it
may be clay is turned into priceless china or
porcelain. We leave the building full of wonder
at the things we have seen, and those particular

manufactures will ever afterwards be invested with a special interest for us, because we have seen with our own eyes how they are produced.

Just in the same way we shall look upon trees in a new light, if we are able in some measure to follow the processes nature is carrying on in them year by year so as to ensure the foliage, flower, and fruit, which minister so much to our pleasure and profit.

The four names we must learn about in order to understand the formation of wood are these. First the outer bark, called epidermis, from two Greek words *epi* upon, and *derma* the skin. *Cortex*, a Latin word meaning bark. Fibro-vascular bundles; this long phrase refers to certain threads or fibres which exist in stems and give them toughness and elasticity. From such fibres in the flax plant we obtain linen, and from the hemp fibres ropes are made. *Fibro* comes from the Latin *fibra*, a thread or fibre; and *vasculum* is Latin for a little vessel; we know the word better, perhaps, in another sense as *vasculum*, the tin box in which botanists place their plant collections.

These thread-like vessels are well called bundles

because they exist in little masses in the substance of the stem.

Most young people know what is called King Charles's Oak in the stem of the brake fern, so plainly seen when it is cut across with a penknife. The dark markings are the ends of the fibro-vascular bundles which happen to resemble an oak tree in form, though some think them more like an eagle with outstretched wings, so the fern is named *Pteris aquilina*, from *aquila*, an eagle.

The fourth word is pith, the white substance in the centre of the stem, which can readily be seen by dividing a piece of elder branch, when the middle will be found full of white pith.

When we have these four parts of the stem clearly in our minds it will be possible to go on with our study and learn about the spaces between, which are filled with different kinds of cells.

The honey-comb formed by bees consists of small cells, little hollow spaces in which they store the honey or bee-food. Woody structure consists largely of cells of various shapes to contain sap and other substances. A beautiful specimen of cell net-work may be obtained by placing a thin

slice of either white or yellow water-lily stem on a
piece of glass and, holding it up to the light, a fine
sort of lace-work will be seen. These are the cells
which convey air and water through the stem up
to the leaves and flowers. Or if we examine a
flower petal with a magnifying glass we shall find
it to be entirely composed of minute cells.

In these little spaces are stored very many

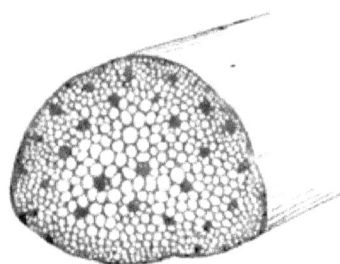

STEM OF YELLOW
WATER-LILY.

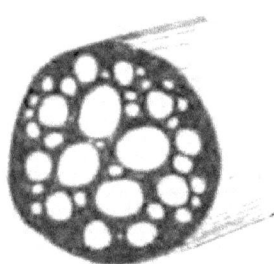

STEM OF WHITE
WATER-LILY.

and very different materials, all necessary to the
growth of a tree; we shall try and learn about
them by degrees; at present we must endeavour
to obtain a clear idea of their structure.

A tree-stem increases in size yearly by the
growth of fresh cells within the outer bark, and
this active increase of tissue is due mainly to what
is called the cambium layer, which is developed
only in the spring and summer and does not exist

in winter; it forms bast, or phloem, on the outer side next the bark, and on the inner side next the pith it creates woody tissue.

Our English lime tree has a layer of fibre beneath the bark which is worth examination; it is the same in character, but not so wide or strong, as the bast which we import from Russia in mats to protect vegetation from frosts. Squirrels are very fond of this soft material; they strip it cleverly off the branches of our lime trees to form a warm lining for their nests.

It is easily found by cutting the outer bark off any small branch of lime within reach, when we can peel off the inner layer of bast, or phloem, as botanists call it.

The phloem from the lace-bark tree of the West Indies is like the finest possible net-work, and is used for many ornamental purposes. *Liber* Latin for the inner rind of a tree, is another term applied to this cell formation.

The study of different forms of woody fibre will be found most interesting.

I obtained one of my best specimens of it by placing a very old Swedish turnip in water for some months until the soft parts had melted away and only the round ball of fibre remained. If any one wishes to follow my example I would suggest placing the turnip and its pan of water in some outhouse where its perfume will not incommode any one. A maid came to me one day with a sad account of a fearful smell which had been noticed for some time in a lumber-room at the top of the house, and very naturally she thought that the plumber should be called in to remedy the evil. I had almost forgotten my interesting skeleton, but in due time I traced the odour to its right cause, and the turnip was banished to a distant spot, where many washings and some soaking in chloride of lime changed it into a really beautiful specimen of woody fibre. I possess now only a quarter of it, for botanists have so earnestly begged for pieces of it that I have been per-suaded to share it with them.

I have sometimes picked up on the seashore old cabbage-stems bleached to a delicate ivory white, forming really beautiful instances of woody fibre. These we can prepare for ourselves, if desired, by soaking the stems in water until they can be

brushed perfectly clean, and then bleached by
mixing a little chloride of lime in water and
letting them soak in it till they are white and
free from odour.

In a manufactory there must of necessity be a
series of windows on the different floors, not only
to let in light but for purposes of ventilation.
Now, the processes of tree-growth are carried on
without light in the stem, but air is necessary, and
it is supplied by means of small apertures called
lenticels. These are not open holes, but are more
like gratings which admit a small amount of air
through loosely-packed cells.

These lenticels are the small brown specks
which may be traced in great numbers on the
young branches of almost any tree. They remain
open through the spring and summer, admitting
the needful air to the interior of the bark, but
when the tree-growth is over for the season, and
air is no longer needed, a layer of cork forms
within the lenticel which entirely shuts it up and
keeps out the wintry cold. Thus it remains sealed
up till, by the growth of the cambium layer in the
following spring, the corky barrier is split open
and air is again admitted.

These lenticels are nature's ventilators, opening
and shutting in this curious way in order that the
manufacture which is going on beneath the bark
may receive from the outer air the various gases
essential to the work which is being carried on
within.

I have said that many and various things are
stored in the stem-cells of trees. It would occupy
too much space to attempt to make anything like
a complete list of the liquids and solids which are
obtained from trees, but I will enumerate a few of
those with which we are familiar from their use-
fulness in every-day life.

Turpentine is obtained from various kinds of
firs—the Scotch fir, larch, and others. Burgundy
pitch from the spruce fir. A kind of tar is also
prepared from Scotch fir and larch. From various
kinds of cinchona we obtain quinine, so valuable
as a remedy for fever. Camphor is a product of a
Chinese tree. Tannin, by which skins are converted into leather, is obtained from the bark of
the oak-tree. A kind of sugar is made from the
sap of the maple, which is largely used in America.
Gum arabic and a great number of gums used in

medicine are produced by foreign trees of various kinds. The interior pith of a West Indian palm tree produces the sago of commerce.

Stems, like every other part of a plant, are to be seen in endless variety when we come to examine them for ourselves. In common garden plants such as the calceolaria and petunia, the consistence is soft, and such stems are known as herbaceous; these generally die down in autumn. Roses and rhododendrons have stems of a harder and more rigid character, and seem to be intermediate between the soft herbaceous stems and tall tree trunks.

If in some country ramble we resolve to make the trunks and bark of trees our study, we shall find much that is interesting and well worthy of observation.[1]

The Lombardy poplar, with its tall bending stem, the graceful willow and the silver birch, contrast strongly with the thick and sturdy trunks of the elm and oak. Even these two differ, the wood of

[1] For instance, I have noticed some curious examples of trees growing together. A Turkey oak and Silver fir in my own grounds are closely united at the base. The fir-seed and the acorn must have germinated in such close proximity that the stems have almost grown into each other. The group of beeches shown in the plate gives another example of interlacing stems and roots.

the elm being short and brittle, whilst that of the

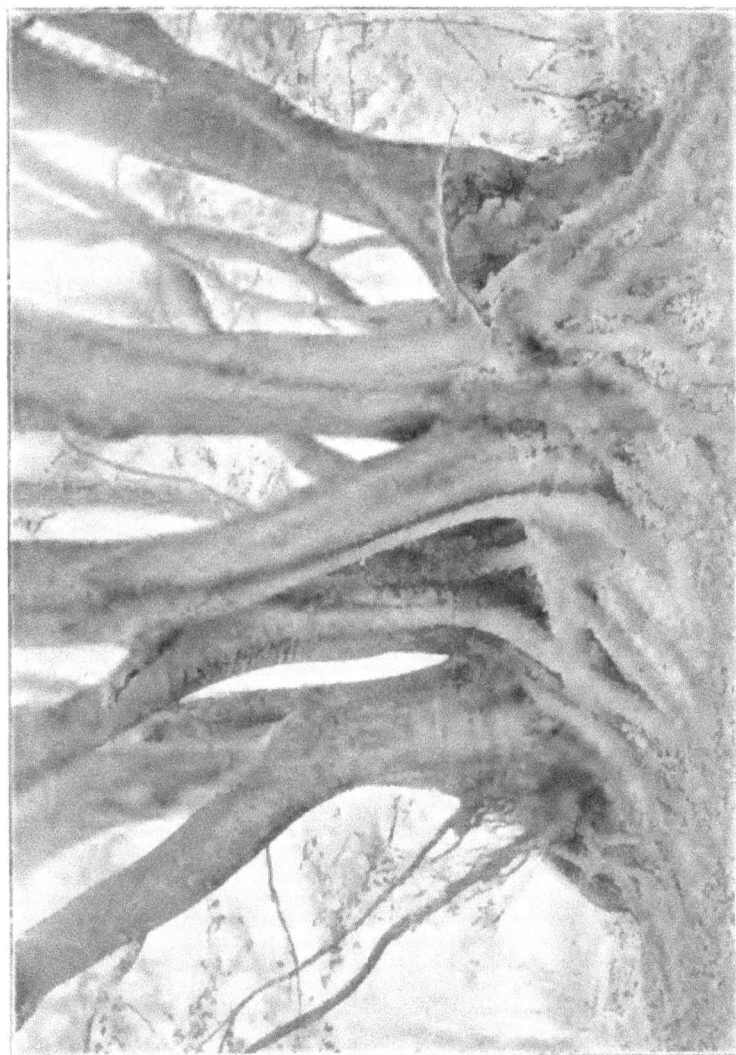

GROUP OF BEECH-TREES WITH INTERLACING STEMS AND ROOTS.

oak is hard and flexible. Again, we may note the
slender drawn-up stems of trees growing thickly

together in a wood, where light and air are in a
measure shut out, and compare them with other
specimens standing in a park in free air and light.
There we see trees, growing as nature intended, with
grand sturdy trunks and well-
developed branches spreading out
on all sides. Lastly, in this chapter, we may note the climbing
stems; these are especially numerous and diversified in their
manner of growth. Almost every
part is modified and adapted to
assist the stem to climb. The
common ivy develops upon the
surface of its stem numerous
rootlets, and by their clasping
nature the ivy is enabled to
ascend the smoothest tree-trunk.

CONVOLVULUS. The hop and the convolvulus
climb by means of their habit of twining around
some rigid stem or twig. Then the peas and
vetches send out little clasping tendrils in the
place of leaflets, whilst that lovely ornament of
the hedges—traveller's joy—climbs by occasionally

using the leaf stalk for a clasping holdfast. Not
less interesting are the plants that climb by means
of their hooks; the common bramble is of this
kind; it scrambles over the hedge in a very enterprising and aggressive manner, while its spines
and hooks effectually prevent it from slipping back.

A very highly developed organ of climbing
is that to be found upon the stems of the
small Virginian creeper (*Ampelopsis Veitchii*).
On the points of its small tendrils we shall discover little globular, crimson-coloured pads, which,
when pressed against a tree or wall, secrete
a kind of vegetable glue. This fixes the
tendril and enables the weak slender stem to
climb upwards. In these instances, as well as
others mentioned earlier in the chapter, we
have again evidences of how wonderfully plants
are adapted to their wants and environment.

Things to observe and collect:—The various
ways in which trees shed their bark; how trees
repair holes in the stem; fibres in flax stem
and in hemp; specimens easily obtained by
sowing linseed and hemp-seed; section of brake-
fern stem; section of elder stem; thin section of

white or yellow water-lily stem; flower petal; piece of bast matting; West Indian lace-bark; turnip and cabbage stalk prepared as specimens of woody fibre; lenticels on various trees; suitable leaves for skeletonising—holly, magnolia, tulip-tree, pear, poplar, aspen, mahonia, plum and maple; suitable capsules—poppy, stramonium, henbane, winter-cherry, campanula, and the calyces of the yellow-rattle.

CHAPTER IV

LEAVES

> " These naked shoots
> Barren as lances, among which the wind
> Makes wintry music, sighing as it goes,
> Shall put their graceful foliage on again,
> And more aspiring, and with ampler spread,
> Shall boast new charms, and more than they have lost.
> Then each, in its peculiar honours clad,
> Shall publish even to the distant eye,
> Its family and tribe." COWPER.

CHAPTER IV

LEAVES

WE have learned in the previous chapters that the roots are the means by which a plant gathers out of the earth the various constituents which are needful to maintain its life.

The leaves have also to do their part in collecting from the air such gases as are required to effect the processes carried on within the substance of the leaf.

The leaf is really the digestive organ of the plant; it feeds, breathes, and gives off in the form of vapour any excess of water not required for its work. For these purposes sunlight and air are necessary.

A leaf consists of a stalk, called a petiole, and the flat green part, which we may call the blade.

99

If we hold a leaf up to the light we see a network of veins, and it is by their help the leaf becomes a broad expansion of tissue, so exposed that it gets the fullest possible benefit from the sunlight and air. This fibrous network gives strength to the leaf, and answers to the bones in animal structure.

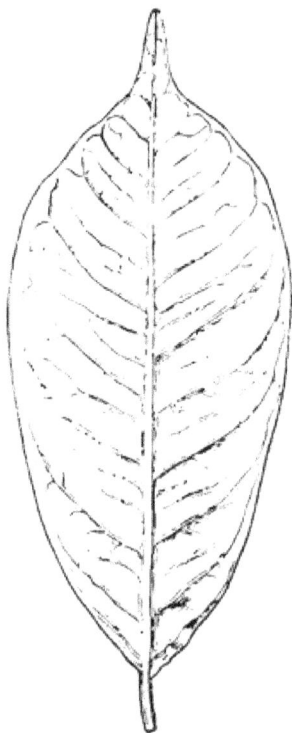

The fibro-vascular bundles, which we see in the stem, go up through the petiole, and branch out in a beautiful and regular manner. We may observe this arrangement very clearly in a skeleton leaf, the mid-rib forming a backbone to the whole structure, while the smaller veins tend off to the edge of the leaf, and then overlap so as to form a system of girders supporting the edge, and preventing the wind from tearing the delicate tissues into shreds.

NETTED VEINS.

The arrangement of leaf network is called

venation, and by a glance at it we can at once
see to which of the great divisions in botany a
plant belongs. If the fibres are straight and run
parallel to each other without being netted, then
we know the leaf is that of a plant which begins
its life with only one seed-leaf; such
are all the species of corn and grass,
bulbs, palm-trees, bananas, and others.

The long name applied to this
division of plants must be explained, as it is a term we cannot
do without, and I must own it looks
formidable until we understand its
meaning.

The first leaf that comes out of a
seed is called a cotyledon, from
kotûle, a cavity, or cup. The Greek
for one is *mónos*, so plants with one
seed-leaf are called monocotyledons.

MONO-
COTYLEDON.

If we sow a date-stone or a few seeds of Indian
corn in moist soil they will grow readily, and
afford us nice little specimens of a one-seed leaf-
plant.

If we see that a leaf has netted veins, then we

know its seed produced two leaves at first,[1] so
plants belonging to this great division are called
called dicotyledons.

DICOTYLEDON

In order to watch the growth of two-leaved
seedlings, we may select a broad bean, or some

[1] The Maranta and a few other plants are exceptions to this rule.

of the seeds out of tamarind jam; either will grow readily in a pot of earth, if it is placed in a sunny window, or near a stove, and kept moist.

Orange and lemon pips may sometimes be found sprouting within the fruit, and either of these seeds will germinate, and form charming little evergreen plants to brighten a town window-ledge.

Now we need not be afraid of those two long words which are used to describe one-leaved and two-leaved seedlings, since we know their meaning, and it will be interesting when we come across some new plant to see to which division it

TAMARIND SEEDLING.

belongs, because knowing that will mean knowing a great deal besides.

All our English trees (with the exception of the firs, which have many seed-leaves) are dicotyledons; they increase their stems from the outside, and are therefore called exogens, and most of our plants belong to this division.

The monocotyledons increase from the centre,

that is to say, the second leaf grows out of the first, and the third leaf and its stem grow out of the sheath of the second leaf, and so on ; and this is the law of their growth, whether they be corn plants or palm-trees. These sheathing leaves and the straight veins will always enable us to recognise a one-seed leaf-plant at sight.

The development of the stem has a marked influence upon the arrangement of the leaves ; these, in such plants as the cyclamen, sundew, or primrose, are said to be radical ; that is, growing from the root. Close observation will reveal the cause to be the non-development of the internodes, the leaves being crowded upon a very short, suppressed stem, and thus we get the beautiful little rosettes we find in the daisy and plantain. When the stem is of greater length the leaves are ranged at definite intervals, the spaces between the leaves (the internodes) varying in length in proportion to the size of the leaf. Small leaves are thus much thicker upon the tree than larger ones. This will readily be seen if we compare a branch of sycamore with one of elm, the former having its large leaves much further apart than the latter.

Then, also, the arrangement of leaves upon the stem (*phyllotaxis*) varies much. If we take a spray of beech we shall find that its buds are placed alternately on either side of the stem, so that the third bud is exactly below the first, and the second bud is in a line with the fourth, and so on. This is also the plan of the elm, hazel, lime, hornbeam, and many other trees. In the alder and white-beam the buds occur in three rows, and in some of the willows in series of eight.

The leaves of the horse-chestnut are borne in pairs on alternate sides of the stem, and this plan is common to a number of plants, especially those of the type of the dead nettle and speedwell.

Quite a distinct arrangement is that to be found in the woodruff and bedstraws, where the leaves are placed in a ring (a whorl) at regular intervals on the stem.

The botanical student should carefully observe the differing methods of leaf arrangement, since, as branches are developed from buds, the varying order in their position must naturally modify the general aspect of a tree, and has also much physiological importance. We shall find that

buds are so placed that each leaf shall receive
its full share of sunlight and air, for it needs this
position in order to enable it to carry out the
wonderful work of assimilation which it has to
perform.

The upper surface of a leaf is covered by a thin
layer of cells, known as the epidermis (or skin);
this does not prevent the light from falling
through, and its outer surface is protected by a
thickening, known as the cuticle. This is of great
use in controlling the escape of moisture, otherwise the leaf would soon shrivel up in a hot sun.
In a young seedling leaf the cuticle is not developed, and it can therefore breathe out moisture
very rapidly ; later on, when the cuticle is formed,
it controls the escape of moisture, which can then
only exude through the under surface of the leaf.

We can easily peel off a portion of the skin
from the under surface of the leaf, and if we place
it in a little water between two pieces of glass
and look at it in a microscope we shall see that
it consists of an extremely thin layer of cells,
with numbers of little openings called stomata
(from the Greek *stoma*, a mouth), answering

somewhat to the lenticels to be found in young
tree-stems, only those are solely for the admission
of air, while these little mouths are to let in and
out not only air, but water, vapour, and oxygen.

These stomata look like little crescent-shaped
slits with a curved cell on either side, and as
they curve more or less, the mouths are opened
or shut as the plant may require. These little
mouths play a very important
part in the economy of the leaf,
and they exist in immense quantities on its under surface.

It has been calculated that a
million stomata exist on a single
leaf of the lime tree. When the
root has taken up more moisture than is required,
then it is the office of these pores, or stomata, in
the leaf to give out this extra water in the form
of vapour, and we can thus see how the action of
leaves must influence climate. If forests are reck-
lessly cut down, the bare country, with no foliage
to throw moisture into the air, may become an
almost barren desert, and again in marshy places,
where the air is too damp, a wise reduction in the

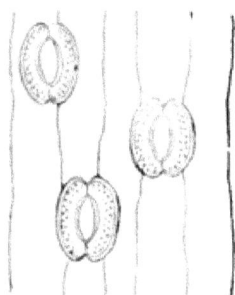

STOMATA.

number of trees may alter the climate to a healthy condition.

Remarkable results have been obtained by planting the Australian gum-tree, *Eucalyptus globulus*; it thrives well in malarious places, and at once produces a marked hygienic change in the air. A Monsieur Gimbert relates that "A farm some twenty miles from Algiers was noted for its pestilential air, and in the spring of 1867, 13,000 eucalyptus trees were planted there, since which time not a single case of fever has occurred.

"The gum-tree grows rapidly and absorbs as much as ten times its weight of water from the soil, and emits camphoraceous antiseptic vapour from its leaves. It is therefore often called the fever-destroying tree."

Experiments have been made to try and find out how much moisture is really given out by leaves. It was found that a sunflower three and a half feet high, with a leaf expanse of over five thousand inches, exhaled one pint of liquid in the course of the day.

No wonder, therefore, that trees tend to make the air damp.

Each stomate leads into air spaces between the
cells, and is thus connected with the interior of the
leaf.

The tissue and cells of a leaf bifacial,[1] can be
understood by reference to the accompanying diagram. Between the upper and under surfaces of
a leaf there is a layer, more or less thick, of soft

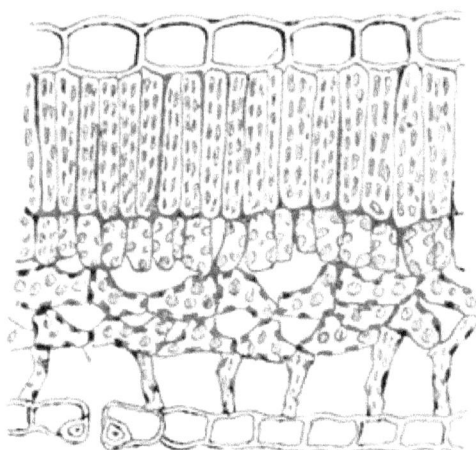

LEAF SECTION.

green tissue known as *mesophyll*, and if we hold a
leaf to the sunlight we shall see the veins travers-
ing this tissue.

The upper part of the mesophyll consists of
elongated cells arranged at right angles to the

[1] That is, a leaf like the beech or sycamore, having an upper and
under surface : vertical leaves, like the iris, have palisade tissue on
both sides.

surface, and placed so evenly parallel to each other
that they have been compared to the pales of a
fence, and are called palisade tissue.   These cells
contain a quantity of the green substance called
chlorophyllon (from *chloros*, green, and *phyllon*, a
leaf), so named because to this bright green substance we owe all the lovely verdure of our woods
and gardens.

Below this palisade tissue is another of quite a
different form, consisting of large spongy cells, and
therefore known as spongy tissue.

In its intercellular spaces are stored those
secretions which make certain herbs, such as
thyme, marjoram, and others so fragrant when
bruised.

The chemical changes which are ever going on
in these various layers, require a constant supply
of the outer air, and this is secured by the little
openings, called stomata, on the under surface of
the leaf, which have been already described ; these
constitute the breathing apparatus of the leaf, for
they open and shut, and regulate the supply of
air into little air chambers, from which it passes
into the structure of the plant.

Before going any further I must try and explain a little about the wonderful substance called protoplasm.[1]

If we have ever watched a potter at work, we know he takes a lump of clay and moulds it according to his purpose, into a rough pot, or a lovely vase ; now protoplasm seems to be just such a foundation material from which the Divine Creator causes animal and vegetable forms to proceed. *First material* seems to me to be a term that actually expresses the meaning of the word protoplasm.

It lines the cell walls of leaves, it is capable of forming fresh cells, it can absorb moisture and other matters, it contracts and expands, it has power of movement, as one may readily see when a portion of a leaf is placed in a microscope, so as to show the grains of bright green chlorophyll circulating in the lining of each little cell.

Learned volumes would be needed to explain the nature of protoplasm, so I must be content with these simple facts about its nature, and proceed to the chemical action going on in leaves.

[1] Greek : *proto*, "first" ; *plasma*, "anything moulded."

In ordinary atmosphere there is a very small quantity of a gas called carbon-dioxide.[1] The leaves absorb this gas from the air, and because there is so little of it, each tree needs to spread out an immense amount of foliage, that it may drink in, by its means, all the carbon-dioxide that can possibly be obtained.

When this gas comes in contact with the chlorophyll in a leaf, one part of the oxygen is set free, and returns to the air in a pure condition, thus making it more healthy for us to breathe; then the carbon and the remaining oxygen combine with water in the leaf cells, and form starch, the leaves retain the carbon, to build up their own structure; it enters indeed so largely into the composition of vegetable substance, that in some cases if we could burn one hundred parts of it, fifty parts of the ashes would prove to be carbon or charcoal.

In a rough sort of way we may see for ourselves how much carbon there is in woody fibre, by lighting an ordinary wooden match and letting it burn

[1] Carbon meaning charcoal, and dioxide meaning two parts oxygen. Sometimes called carbonic-acid gas.

itself out ; the black portion that remains will be a piece of charcoal not very much smaller than the original match.

Of course, in the process of burning, the match has lost the resin, and other organic substances which were stored up in the cells of the wood ; all these have passed into the air, and only the carbon remains. If, however, instead of this slow manner of combustion, we had set light to a whole box of matches so that it burnt fiercely, the flame would have been strong enough to consume the charcoal, and nothing would then be left but mineral ashes.

When charcoal burners are at work in a forest, we may see them making a stack of wood, which they cover with a thick layer of clay so that the wood may burn away very slowly ; in this case the charcoal will be left in the same way as when we burnt the single match.

As long as the upper side of leaves are soaking in the sunlight, starch is being formed, as I have described ; but during the night the starch thus formed is dissolved, and passing through the leaf fibres finds its way into every part of the plant,

8

either to be used in forming new tissue, or else to be stored up for future use.

The net-work of veins act as a service of tiny pipes, to convey the liquids up and down the petiole (leaf-stalk).

Generally the water given off from the stomata is in the form of vapour; but in some plants drops of water exude from the apex or point of the leaf through the water pores. In *Saxifraga crustata*, there are pores round the edges of the leaves, through which water, highly charged with lime and other salts, passes out, and as it evaporates a white deposit of lime remains which is quite visible in the form of a frosted edging to the leaves.

There is an American plant called the jewel-weed, which shows to perfection this power of distilling drops of water. I will quote a short description of its appearance at night-fall.

"Upon the approach of twilight, each leaf droops as if wilted, and from the notches along the edge, the crystal beads begin to grow until its border is hung full with its gems. It is Aladdin's lantern that you see among a bed of these succulent pale

green plants, for the spectacle is like dream-
land." [1]

A very similar effect may be observed if we
visit a plant of Lady's-mantle (*Alchemilla vulgaris*)
at early morning after a warm dewless night : each
leaf will be found beautifully decked with dew-
drops at equal distances round the edge of the
leaves where the pores have exuded the moisture
with which they are charged.

Nasturtium and fuchsia may also be examined
for this purpose and will show exudation from
their leaf pores.

If a small quantity of wheat is grown in some
cocoa fibre, it will illustrate this power of giving
off water, for when the little blades are a few
inches high, they will be found each morning
tipped with a large dewdrop, the result of exudation during the night.

In countries where the sun is intensely hot, if
the leaves of trees were to be exposed to its full
power, they would probably wither, and vegetation
would perish.

Against this danger some trees are enabled to

[1] " Sharp Eyes," by Wm. Hamilton Gibson.

make special provision, by changing the form of
their leaves, and their mode of hanging on the
branch.   In Australia, for instance, where the sun
is almost vertical, the acacias and eucalyptus trees,

YOUNG SHOOT OF EUCALYPTUS.

instead of holding their leaves flat or horizontally
as trees do in England, so that they may catch
every ray of sunlight, avoid the heat as much as
possible, by holding them edgeways to the light.

MATURE FORM OF EUCALYPTUS LEAVES.

While eucalyptus trees are young, and partially
shaded by surrounding vegetation, their leaves are
flat and oval, and English seedlings of this tree
usually retain such leaves from five to ten years,
our climate not being hot enough to require the
mature form of leaf which hangs vertically, and is
of an entirely different form.

Reference to the plates will show a young shoot
of *Eucalyptus globulus* and a branch of the older
leaves, with their edges only exposed to sunlight.

The curved sickle-shaped leaves of the eucalyptus afford very little shade to the traveller in
Australia for this reason, that only fine inter-
crossing lines of shadow are seen on the ground.
To make this clear, let my readers take a sheet
of notepaper out of doors on a sunny day and
hold it perfectly flat, so as to expose it to all the
sunlight it can receive upon its surface, as if it
were a growing beech-leaf, and it will throw a
large shadow on the ground. Then hold it edgeways to the sun, and it will form the kind of thin
line of shadow that would be cast by a mature
eucalyptus leaf.

Preparation for the fall of the leaf begins in

spring, when a fine line or ridge may be traced just below the junction of the leaf with the stem This dark line is in reality a thin layer of cork, which, during the summer months, continues to grow inwards to form in due time a covering for the bare place on the stem that will be left when the leaf falls off; this is called the leaf-scar.

It is interesting to watch this line, growing more and more visible as the year goes on.

Another curious fact is, that some of the starch which the leaves have been making during the summer, becomes stored up in autumn at the base of the leaf-stalk, so as to afford nourishment to the bud which will arise out of the axil of the leaf. When a weak solution of iodine is applied to it, this starch turns blue, and in this way its presence can be ascertained.

The fall of the leaf appears to take place mainly because the starch has the effect of softening the cells of the leaf-stalk ; as it dries up it loses its hold of the twig, and either the wind or a slight frost will suffice to bring the leaves down to the ground in showers.

Another reason for their fall is, that their year's work is done. Like good servants, they have been hard at work all through the summer and autumn months, taking in stores of nourishment for the benefit of the tree, and giving out volumes of oxygen, so helpful for the maintenance of human life. They have secured and laid up sufficient nutriment for the development of the next year's buds, and having done this, their special office being at an end, they fall beneath the tree to become leaf-mould, which, in its turn when fully decayed, will yield nourishing elements to be carried by the winter and spring rains to the tree roots.

I might add many more useful objects which we owe to trees, and I commend it to my young readers as an instructive study to try and make out a complete list of the useful products of our English trees. I imagine we do not yet know all that might be obtained from them, new discoveries continue to reveal their value in the way of medicines; for instance, the crystals of the willow (called salicine) are now frequently prescribed as a remedy for rheumatism. Euonomine

and many others might be included amongst the
valuable gifts which nature has stored in the cells
of tree-stems.

Specimens to be obtained: Leaves with straight
veins, such as grass or corn, hyacinth, narcissus, or
any bulbous plants ; leaves with netted veins, such
as oak, ivy, vine, &c. ; monocotyledon seedlings ;
dicotyledon seedlings. Leaf-skin to be examined
through a microscope, in order to see stomata,
chlorophyll, network, and cells. Examine water-
pores in leaves when exuding moisture. Observe
shadows thrown by leaves held flat and edgeways
to the sun. Compare young and old eucalyptus
leaves. Observe line of cork below leaf-scars.

Leaves can be made into beautiful skeletons by soaking a good
many together in a pan of soft water until the upper and under
surfaces of the leaves are sufficiently decayed to be easily removed
by a soft brush ; the fibre which is left can then be bleached with
chloride of lime. When mounted with fine wire these skeleton
leaves form an interesting group to place under a glass shade.

CHAPTER V

BUDS

" Oh ! who can speak the joys of spring's young morn,
When wood and pasture open on his view,
When tender green buds blush upon the thorn
And the first primrose dips its leaves in dew."

CLARE.

124

CHAPTER V

HOW we watch for the buds as tokens of the coming spring! We delight to see them daily growing larger, and opening out their leafy treasures to the sun.

The re-clothing of the trees has always an element of wonder about it; the change is truly a resurrection; a few days of warm sunshine and gentle rain, and then the dry, dead-looking branches begin to bud and blossom as if by a miracle.

We may, however, trace the secret of this sudden change, if we look back to the processes Nature was carrying on during the previous summer, and we may learn from her many a useful lesson of foresight and preparation.

125

If during the summer we lift up the branch of any deciduous tree and search amongst the leaves, we shall find that the buds for the following year are already there, waiting to be developed in due time. When the leaves turn dry and sere, they fall off and leave the buds to be hardened and matured by the rain, snow, and frost of winter.

Certain species of Turkey oak, young beeches, hornbeams in hedges, and other trees, appear to retain their leaves, to some extent, as if to protect the buds until the rising sap in spring loosens their hold upon the branches, and makes them fall off.

The plane-tree appears to be an exception to most trees in the curious protection it affords its young buds. If we search for these in the summer or autumn, they are not to be found, for the leaf-stalk is so swollen and hollow at the base as to enclose the bud within it; even when the leaf falls off, the bud is covered by a tough outer case coated with resin, and within are many fur-lined scales. When these are removed we see the tiny leaves wrapped in silky coverings, and when the warmth of spring enables them to expand, their under surfaces have such a thick coat of down that the plane is known in some countries as the cotton-tree. From its fruit being in the form of

OAK IN WINTER.

hanging circular balls, its name in America is the button-wood. The need of this special protection against cold is shown by the fact that if severe frost returns after the leaves have expanded, they frequently shrivel and perish. Some Japanese maples have the same arrangement of hollow leaf-stalks to contain the buds.

When buds are situated at the end of a branch they are called terminal, and their office is to increase the length of the branch.

When they grow in the axil of a leaf (that is, where the leaf-stalk joins the stem) then they are called axillary, and as they grow out and form fresh stems and leaves, the branch broadens on either side.

Seeing that the branches of a tree thus spring from the buds, it follows that the position and development of the buds upon the stem, as we tried to show in the last chapter, have much to do with the ultimate shape of the tree. The development of the axil buds, as well as of the terminal bud, gives rise to a branched tree like the oak ; these buds, however, are often erratic, and in some trees the terminal bud of the shoot is often suppressed and the axil buds grow with extra vigour, whilst in other instances the terminal bud grows strongly and the axil buds either grow feebly or are altogether suppressed. In the bamboos, palms,

and sugar-cane we get good examples of this
terminal bud-growth, the axillary buds being
suppressed ; the suckers that grow from the axils
of the lower leaves of the palm are often evidence
of the presence of axillary buds, although they

OAK IN SUMMER.

are, as a rule, dormant.   We are all more or less
familiar with the character of ordinary forest trees,
the rounded outline of the oak, the slender sprays
of the birch, the spreading branches of the beech,

9

but perhaps we may not have remarked how much these variations of form are due to the position of the buds upon the branches. We will suppose that on a winter's day we are looking at the tracery of some elm-branches against the sky; the form of each branch shows that the terminal bud in this tree usually ceases to grow, and allows the lateral shoots to increase in length, and take its place; this produces short, twiggy branches, and a stem which makes a tall tree rather than a wide-spreading one. The horse-chestnut, again, produces its flowers in the terminal buds; this arrests their growth, and side shoots have to grow on instead, thus usually giving height rather than breadth to the tree. We may note the differing outline of the willow, birch, and many others where the terminal buds do not cease to grow, but each year continue to add to the breadth as well as the height of the tree.

In pine-trees the buds are produced at the ends of the branches, and several shoots proceed from one bud.

The spiral arrangement of leaves is well seen in a young coniferous shoot, also in the flower-bud, and especially in the fir-cone itself, in which an ever-varying double spiral can be traced.

Loudon remarks, " The perfection of a fir consists in height rather than in lateral expansion ; buds are produced very sparingly and nearly always at the extremities of the shoots. Provision is thus made for the upward growth of the tree more than for side expansion."

When we speak of a coniferous shrub having lost its leader, we mean that the terminal bud on the topmost shoot having been broken off, one or more of the lower branches must rise up and take its place, and thus lateral buds in time become terminal and grow upright instead of sideways.

A silver fir, that I have been observing for years past, bears such a crop of heavy cones on its slender upper branches that the leader is almost invariably broken off by the weight, and the lateral shoots have to take its place, to the great detriment of the central stem, which is twisted and curved out of shape by the efforts the tree makes to repair its terminal shoot.

In other trees, again, the unfolding of all the buds is nearly simultaneous, but in the fir tribe the bud which terminates the summit of the tree and is destined to form its leading shoot and increase its height is developed last ; this delay seems a provision of nature for the safety of the most important shoot which the tree can produce, ensuring its height rather than its breadth, and the production of timber by the preservation of its permanent trunk rather than by its temporary branches.

If a willow is deprived of the upper part of its stem and so made a pollard tree, it develops a bushy head of small stems which spring from the other buds thrown out to repair the loss of the central stem. This pollarding is often resorted to in order to obtain wood of the right kind for basket-making, and young ash trees are thus treated, so that slender rods suitable for hop-poles and tool-handles may spring from the lopped stem.

When buds are found growing on any other part of a plant except those just mentioned, they are called adventitious buds. These may be found growing on the edges of the leaves of the marsh tway-blade ; they also spring out of the flat surface of the fronds of the viviparous fern.

Under favourable conditions every part of a plant will produce buds, and, taking advantage of this fact, florists increase their stock of succulent plants by putting the leaves on a wet surface, which induces them quickly to send out buds and roots. Such plants as begonias, gloxinias, hoyas, and sedums are readily increased by this mode of propagation. Underground stems will often send out buds, and they produce the underwood from the stumps of fallen trees.

We are all familiar with the suckers of trees which spring up in our lawns and gravel paths often many yards away from the parent tree ; these all arise from active buds on underground stems. Gardeners are always careful to prune away such growths at the base of their wall-fruit-trees, since they rob their valuable peaches, nectarines, and apricots of strength and nourishment. These well-named " suckers" spring from the common stock upon which the choice fruit-trees were grafted, as one may see by gathering a leaf from a sucker and comparing it with a peach or nectarine leaf.

On the oak, chestnut, lime, beech, and other trees there are sometimes to be found dormant buds in the form of rounded knobs covered with bark and increasing in size with the growth of the tree; these, in the event of other buds perishing, will start into active growth and do their part in preserving the life of the tree.

Such woody balls when found on the oak are worth examination, as when divested of their bark they show exquisite structure of woody fibre.

The small bulbils we find in the axils of lily stems, on the cuckoo-flower, on *Dentaria bulbifera*, and on some species of Allium, are all adventitious buds, which drop off in due time and become young plants.

They are in many respects similar to bulbs, and if we cut one in half and compare it with a divided hyacinth we shall see that they both consist of over-lapping scales. In the onion these scales are fleshy and succulent, but in most tree buds they are dry, hard membranes.

The pear and magnolia buds are secured against wintry cold by woolly linings to the scales, and in the horse-chestnut they are covered

HORSE-CHESTNUT.

with a kind of resin which renders them impervious to moisture.

It requires a careful use of the microscope to trace all that a bud contains; I will therefore quote the words of a German naturalist who dissected a horse-chestnut bud gathered in winter, and found that it contained sixty flowers. It would be interesting to select a terminal flower-bud of this tree; by taking it carefully to pieces one might, with patience and using a powerful lens (or a microscope if one is available), see for ourselves a good deal of what the writer describes :—

"Having removed the outer scales, seventeen in number, cemented together by a gummy substance to render the bud waterpoof, I discovered four leaves surrounding a spike of flowers, so clearly visible when magnified that I not only counted the number of flowers, but could discern the pollen on the stamens."

The winter covering of a bud, both the inner and outer scales, are only a temporary protection in order to keep out moisture and keep in warmth, so that as the sun begins to gain power, the gummy covering of the bud melts and yields to the expanding pressure from within, when one

after another the sticky scales fall off, showing the
young leaves with their soft woolly surfaces;
these leaves rapidly unfold and hang droopingly
until the midribs gain strength enough to hold
them upright.

Evelyn remarked that, " As soon as the leading
shoot of the horse-chestnut has come out of the
bud, it continues to grow so fast as to be able to
form its whole summer's shoot, sometimes eighteen
inches long, in about three weeks. After this it
grows but little more in length, only thickens,
becomes strong and woody, and forms the buds
for next year's shoot."

Buds have always been to me a most interesting subject of study; there is much variety of
character in them, and to those who observe them
closely they reveal in the autumn and winter
what the tree is purposing to do in the following
season.

A beech-tree on my lawn bears its nuts only
every second or third summer, and in the previous
autumn I can always tell whether the squirrels are
likely to be well off for food in the coming year,
by observing the size and shape of the buds.

Those which contain the flowers are round and bulky, whilst the leaf-buds are long and slender.

Embryo flowers are disposed in the buds in different ways. The wood-sorrel is rolled into a spiral, rose-petals are placed one within the other,

YOUNG BEECH.

the pink is folded in five divisions, and others are pleated and fluted into the smallest possible space. Perhaps of all others the bud of the great Oriental poppy is the best example of exquisite packing. Early on a summer's morning you may see its huge green hairy bud at the end of a stem several feet in length, and whilst you are looking the sepals or calyx leaves suddenly divide and fall off, the mass of vivid scarlet petals crumpled into a thousand folds begins to open out, and before long the glorious flower, which is often as much as seven inches across, holds itself erect in majestic beauty.

Those who possess a tulip-tree will find its opening buds reward examination. The leaves are folded in half and bent double, a pair of leaf-scales enclosing each of the true leaves. One may unpack the entire bud until we come to leaves almost too minute to be discerned.

Young sycamore-trees often have buds of large size and brilliant crimson colour; the foldings of their leaves are very intricate, and form an interesting contrast to those of the tulip-tree. Hart's-tongue fern, arum, and pear leaves afford three very remarkable modes of folding in the bud.

Another point of character in buds is of considerable importance to the horticulturist, namely, the fact that in some cases the value of the flowers produced varies with the position of the buds. For

instance, the blossoms produced from the crown-
bud [1] of certain chrysanthemums are poor and pale

PEAR LEAF.

UNFOLDING ARUM
LEAF.

UNFOLDING LEAVES
OF HART'S-TONGUE
FERN.

in colour compared with those grown on the side

[1] The uppermost bud of the central shoot.

shoots ; the latter are therefore retained and
fostered, so that from them flowers of the finest
description may be obtained.

In cultivating fruit-trees it is found needful not
only to prune away redundant branches which
bear leaves only, but also where strong woody
roots are promoting the growth of leaf-buds, they
also have to be pruned, so that the check thus
given to the growth of the tree may result in the
formation of fibrous roots, which will tend to the
production of flower-buds and a resulting crop of
fruit.

I have often observed that the transplantation
of trees leads to their throwing out flowers in the
succeeding year. This was notably the case with
an avenue of deodars which had overgrown my
carriage drive ; they received a considerable check
in being transplanted, but in the following year
their branches were covered with male catkins and
some few cones succeeded.

For this reason the removal of fruit-trees is not
unfrequently resorted to, as a means of inducing
fruit-bearing.

So much vigour is stored up in the bud, that it

will bear being removed from one tree and inserted in the stem of another, within which it will grow and become a part of the living tree. This is one of the means by which we have obtained such an infinite variety of roses; the buds from choice species being readily made to grow upon strong briar stocks, and thus one may also see roses of several different colours blossoming on the same stem. Choice varieties of fruit-trees are cultivated in the same way by means of buds inserted in the bark.

Having observed how flowers are arranged in the bud, we may go on to dissect incipient leaves and learn how they are placed.[1] We shall find that the frond of the hart's-tongue fern is rolled up from the tip, the arum gracefully curved lengthwise. Pear leaves are rolled from side to side towards the middle, and so is the primrose, but the reverse way. Beautiful examples of curled leaves may also be seen in the water-lily and banana.

In grasses the first leaves are equivalent to budscales, and protect those which continue to grow from the centre, each one sheathing out of the

[1] Venation.

previous leaf after the manner of monocotyledons.[1]

The colouring of buds is one of the lovely features of spring. Seen against the blue of the sky, the coral red of the lime, sycamore, and Japanese maple buds, cannot be passed by without notice. The whitebeam has a beauty of its own for its buds are large and white with downy coverings, giving promise of the future leaves which are so light-coloured underneath, that the effect when they are blown aside by the wind is curious and beautiful. The Germans call it *mehl-baum* or mealtree, from its whitish downy leaves.

The variegated vine, sometimes seen in greenhouses, has exquisite buds of pinkish crimson, with bright yellow stipules. By way of contrast, I once placed some sprays of it in a glass with twigs of purple hazel which are of a deep claret brown : they were not only opposite in colour, but curiously different in habit, the vine holding its bud erect, and the hazel as persistently drooping. These variations lead me again to remark that, to a close observer, buds will be found to differ much in

[1] One seed leaf plants.

character and to be well worthy of close attention.

I will mention some trees whose buds are specially remarkable for beauty of form whilst unfolding.
The mountain ash has very graceful leaves when
just emerging from the bud; they show on their
upper and under surfaces two distinct shades of
green.

The unfolding weigelia buds are extremely
pretty in shape, the leaves being pointed and
delicately curled.

I need hardly mention the beech; nothing can
be more exquisite than a spray of its opening
buds with their silky fringed young leaves and
crimson leaf-scales. I look forward every spring
to the joy of watching the unfolding of these
caskets.

A warm shower or two and some sunny days
cause them to expand with a rapidity which seems
magical, and one almost regrets to find the beauty
of the buds in their early stages so quickly passing
away. The ash attracts notice by its jet-black
buds, and the wayfaring tree by the delicate venation of its young leaves.

I cannot refrain from mentioning another beautiful effect arising from young buds in the case of a *Picea nobilis glauca*, which long name simply means a sea-green silver-fir, standing on our lawn. In the summer its terminal buds are a very pale sea-green, and as they grow and are seen against the dark green of the rest of the foliage the effect is very curious, as though each branch had become tipped with frosted silver.

The soft silky buds of the willow, and especially those of the low growing sallow which are gathered as "palm" for church decoration, are amongst the welcome signs of early spring. The sallow has its male blossoms on one tree, but not far away we shall find the female tree bearing

BUDS OF WAYFARING TREE.

the flowers which will eventually produce the seeds. We may therefore seek for three kinds of buds, those which produce the flowers on each tree, and

the others which will clothe the tree with leaves
when the blossoms are over.

This chapter shows us how much there is to
instruct the student of nature during the winter as
well as the summer months.

I have but indicated a very few out of the many
lines of study which may be taken up; one could
write essay after essay upon the growth of a single
hedgerow, but all I can hope to do in simple
chapters of this kind is to throw out hints and
indications, and trust that my young readers may
find their interest sufficiently excited by what they
have read, to lead them on to fuller, deeper study of
each point touched upon.

Nature is an inexhaustible storehouse of
wonders, and the further we explore, the more
our eyes are opened to see the vistas that lie
before us, branching out in various directions.

This special path of botanical study is one that,
more or less, can be pursued at intervals, as opportunity may offer through life, and as it adds much
pleasure to leisure hours, I specially commend it
to my young readers.

Specimens to be obtained and compared with

the descriptions in this chapter. Search for buds
in summer ; plane-tree buds ; Japanese maples ;
terminal and axillary buds ; observe shape and
outline of trees ; buds of coniferous trees ; fir cone ;
fir-tree that has lost its leading shoot ; pollard
willow, and other trees, buds on marsh tway-blade,
and viviparous fern ; buds on underground stems ;
suckers from wall-fruit trees ; dormant buds or
knobs on tree-stems ; bulbils on lily, dentaria
and allium ; horse-chestnut terminal bud ; observe
leading shoot of horse-chestnut in early summer ;
flower and leaf buds on beech ; various flower-
buds ; Oriental poppy ; tulip-tree buds ; various
leaf-buds unfolding ; colouring of Japanese maple,
lime, and sycamore buds ; whitebeam ; variegated
vine ; purple hazel ; mountain ash ; spray of beech
buds ; ash buds ; *Picea nobilis glauca* ; willow and
sallow, male and female flower buds and leaf
buds ; bamboo.

CHAPTER VI

FLOWERS

" Your voiceless lips, O flowers, are living preachers :
    Each cup a pulpit, every leaf a book
    Supplying to the fancy numerous teachers
            From loneliest nook "

HORACE SMITH.

CHAPTER VI

FLOWERS

INSTEAD of looking at flowers as bright and beautiful objects made to be a source of continual delight in our daily lives, though such they truly are, we will rather now, for purposes of study, consider them as the means by which the plant carries out the purpose of its creation, namely, to perfect its seed and thus perpetuate its species.

In the life-history of shrubs, trees, and plants we find this is their one aim, and that everything else is subservient to it.

The stamens and pistil being of essential importance in forming the seed, we find them placed for safety in the centre of the flower; folded round them are the petals or coloured parts of the flower, and outside these again are the green sepals, or leaves of the calyx.

These two sets of enfolding leaves are called "floral envelopes," because they fold over and protect the central organs, the stamens and the pistil.

We will select a buttercup as a type, and taking it to pieces we will try to learn the names and uses of its various parts.

The outside is a greenish-yellow cup which is called the calyx.

The divisions of this little green cup are called sepals, and their office is to protect the five bright yellow leaves within, which are called petals when we speak of them singly, but, taken all together, form the corolla.

In the buttercups the petals are all separate, but if we look at a primrose we shall see that the corolla is in one piece, united in a tube; so also is the calyx.

The botanical term for a corolla thus formed is *gamopetalous*, a long word but easily understood when we know that *gamos* means united; a flower with petals in one instead of many divisions is

more easily referred to by this word than if we
had each time to express it by a sentence.

Gathering a newly-opened flower, we can see at
a glance that the sepals are placed quite below the
central green organs of the flower, and that they
are in no way influenced by the petals ; we also see
that the petals are entirely separate from the other

PRIMROSE.

parts of the flower, and we learn, as the result of
our examination, that the parts of the buttercup
are *free*. To express this botanically we prefix
the word "poly" to the words sepals and petals,
and so we get *polysepalous*, meaning that the sepals
are quite free and distinct, and *polypetalous* referring
to the same condition of the petals.

Now, having removed the petals and sepals, we

can proceed to study the other parts of the flower.

First we find a great number of little yellowish stalks tipped with tiny pouches; these are the stamens, and in the little pouches (anthers) the yellow powder termed pollen is developed. We will carefully take away these stamens, and note in so doing that they are all distinct and all sprung below the green central part. Like the sepals and petals, we find the stamens are free and uninfluenced by the other parts. If we again compare this with a primrose-flower we shall find a difference; the stamens of the primrose spring from the petals and are therefore called *epipetalous* (*epi* upon, a petal). Again in the sweet-pea or scarlet runner we find the stalks of the stamens are all joined together. We now have left upon the flower-stalk the little central green parts previously mentioned; there are quite a number of them; each one is distinct from its neighbour and is free. These bodies are known as carpels, they are large at one end and taper to a curved point at the other, the broad end being attached to the stem. Collectively these carpels constitute the *pistil*, and because the carpels are apart and free it is said to be *apocarpous*.

The flower of the little woodsorrel (*Oxalis acetosella*) will help us to understand better the arrangement of the carpels. If we take away the sepals, petals and stamens, we shall have only the carpels left, and these are five in number. They are in the same position as those of the buttercup, but they are not separate, they are joined by their inner surfaces. We can plainly see that this is the case, since each carpel is distinctly outlined and there are five little tapering ends (stigmas). The pistil in this case is said to be *syncarpous*.

Names are given to express some quality, and they often draw our attention to interesting facts about the plant's mode of growth or the place where it is found ; for instance, the pretty blue nemophila is so called from *nemos*, a grove, and *philo*, I love, because it delights in shady places.

Geranium is derived from *geranos*, a crane, because the fruit of some of the species resemble the beak of that bird.

Some plants are named after famous botanists, as *Linnæa* after Linnæus.

Others derive their names from their mode of growth, as stone-crop, which is called sedum, from *sedo*, I sit, the plant having scarcely any stalk, and sitting, as it were, on walls and rocks.

These instances will show that it is well worth while to study names and learn their meanings, as they often throw so much light upon the history of a plant.

In the flowers of bulbous plants we find that the calyx and petal leaves are frequently alike in colour and texture ; in that case the three sepals and three petals, of which they usually consist, are spoken of as a perianth.

In looking at the brilliant colouring of a flower we should hardly imagine that the petals have to some extent the nature of leaves, and under certain conditions may be changed to the green colour and form of ordinary leaves.

In very wet seasons we may sometimes find rose-buds with the sepals of the calyx developed into perfect green leaves. The floral envelopes therefore possess the nature of true leaves.

The brilliant scarlet so-called flowers of the poinsettia are really coloured bracts, the true

POINSETTIA.

flower being the small inconspicuous blossom in the centre.

In the chapter on leaves we saw that bracts are those small imperfectly-shaped leaves in the axils of which flowers are placed. They are usually green, but may be also brilliantly tinted as in the mauve-coloured Bougainvillia, the bright violet spikes of the *Salvia Hormincum*, and also pure white as in the spathe of the arum.

By special cultivation flowers can be made double, for excess of nourishment will cause the plant to multiply its petals. Instead of the five pink petals of the wild rose we find one of our garden roses bearing as many as eighty or a hundred petals.

Double flowers but rarely produce seeds, because the stamens and pistil have been turned into petals, and as there is no need to attract insects for fertilising purposes, there is no secretion of honey, and therefore we scarcely ever see honey-bees in double flowers; they are wise enough to know that their visits to them would be in vain.

In composite flowers such as asters and sunflowers the change, when they are double, occurs in several ways.

The centre may become filled with florets similar to those in the outside ring, or the florets in the middle may become larger or of a different colour.

These various changes may be readily observed in the cultivated chrysanthemums, in which every form and variety of flowering can be traced.

When the pollen has reached the pistil the flower begins to fade, because its end has been attained; nature, however, has such variety in even the smallest of her operations that the passing away of a flower is accomplished in different ways. In the primrose the corolla withers and drops to the ground. The flower of the spiderwort, one of our common garden plants, becomes pulpy as it fades, in this way resembling the pine-apple plant, the flower of which eventually becomes the luscious succulent fruit.

The poppy is proverbial for its fleeting petals, which scarcely last more than a few hours, a passing wind soon scattering them far and wide.

> " Pleasures are like poppies spread,
> You seize the flower, its bloom is shed !
> Or like the snow-fall in the river,
> A moment white—then melts for ever." (BURNS.)

Some flowers, as the hydrangea, have persistent

petals, which simply lose their brilliant tints and become tough and brown.

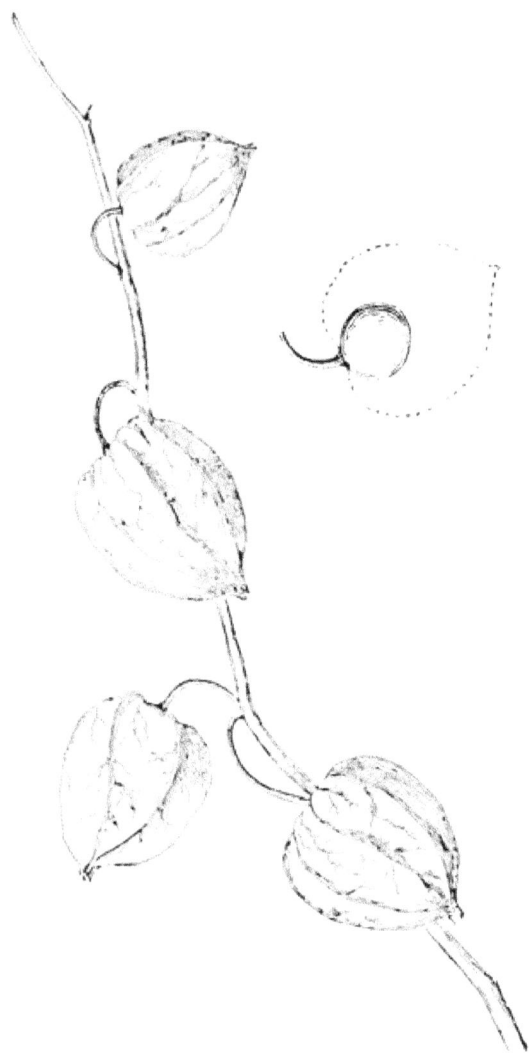

WINTER CHERRY.

The calyx of the physalis or winter cherry

continues to grow after the flowers are fertilised until the round balloon-like bag is formed in which the fruit is enclosed.

We will now examine the parts of a flower separately, beginning with the calyx.

In the buttercup the calyx consists of one whorl or ring of five sepals.

In the strawberry there are two whorls of sepals, and in the cotton plant there are three whorls forming its green calyx.

There are also variations in the mode of flower expansion.

As a poppy-bud opens it detaches its calyx from the stem, and the sepals fall off (the calyx is therefore called caducous, a term which means ready to drop off).

Many flowers retain the calyx until the petals wither and it falls off with them. It is then called a deciduous calyx.

Others again have a permanent calyx, so that when, as in the primrose, the corolla withers and drops off, the sepals close over the seed-vessel and protect it until the seeds are matured; this would be called botanically a persistent calyx.

The best way to learn the names of the different parts of a flower is to pull it carefully to pieces and arrange the separate organs on a thin card. They can be tacked on to the card with a stitch or two of fine thread, and when the lesson is over, if the card is placed between sheets of blotting-paper under a weight, the flower dissections will dry and be useful for reference later on.

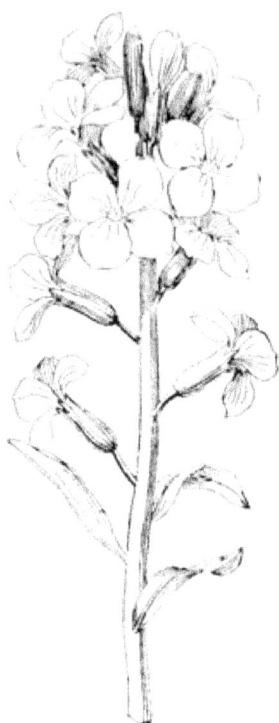

Each separate part of the flower should have its name neatly written beneath it, so that when a good many different flowers have been thus dissected they may be compared and the variations in form and position duly noted.

WALLFLOWER

A wallflower will be a good subject for our dissection.

At the back of the petals we first take off the calyx, which consists of four divisions called sepals. We then pull off the four yellow petals,

and as they are placed in the form of a cross it
shows that this plant is a crucifer, or cross-bearer,
one of a very large natural order, *Cruciferæ*,[1] none
of which are poisonous and very many are useful
food-plants, such as cabbage, turnip, watercress,
and cauliflower. Now there remain six stamens
- four long and two shorter ones; these last rise
outside of and alternate with two nectaries or
honey-glands.

The stems of the stamens are called filaments,
from *filum*, a thread; and the upper part, containing
yellow powder, is called the anther, the proper
name for the powder itself being pollen.

In the centre of the flower is the pistil, the lower
part of which is the ovary, the part of a flower
which contains the ovules, and is so named from
ovum, an egg.

The stem part of the pistil is called the style
and the top of it is the stigma.

Such simple words as I have given must be
learned, else we cannot understand botanical
descriptions, and if this page is studied whilst we
have the flowers in our hands it will not be difficult

[1] All cross-shaped flowers do not, however, belong to this order.

to identify each separate organ ; when these are once arranged on a card with the name of each part written beneath it, we shall have attained some very useful information ready for future study.

In the buttercup flower all the five petals are the same size and shape ; therefore, like hundreds of other evenly-formed blossoms, it would be described as " regular " ; but if we take a sweet-pea, balsam, or monkshood-flower and examine its separate petals, we shall find they vary very much in form, and they are known as " irregular " flowers.

The sweet-pea is a type of a large order of plants producing what are called butterfly-shaped flowers, and *papilio* being Latin for a butterfly, they are therefore called papilionaceous flowers. If we learn clearly about the various parts of such a flower we shall henceforth be able to recognise it at a glance.

In the sweet-pea we find a broad petal at the back of the flower which is called the standard, beneath it are the two side petals called wings, and within them is the keel, so named because it is

shaped like the bottom of a ship. Within the keel
lie the stamens and pistil—the most important
parts of the flower, and to protect them from injury
the standard is so formed as to catch the wind like
a sail and turn the blossom round so that this
broad petal shelters the keel from rain.

SWEET-PEA.

In our next ramble out of doors it will be well
to try and gather all the specimens we can find of
this order of plants. If it be in summer or autumn
we shall soon collect a handful of these butterfly-
shaped flowers.

On a common we shall find broom, furze, rest-

harrow, vetches, tares, trefoil, clover, saintfoin, and
other plants. In the garden and greenhouse we
shall see many more species belonging to this
class.

Having shown the difference between a regular
and irregular flower, we will now proceed to notice
how irregularity is caused.

If we pull off one of the buttercup petals and
look at the base of it, we shall see a small pouch
which contains honey ; it is called a nectary or
honey gland, and the position of this gland has
much to do with the shape of the flower.

As each petal of the buttercup has a nectary at
its base it follows that, all the petals being the same
size and shape, the flower is perfectly regular—
like a small golden cup. Now in other flowers we
shall find the nectary very large and confined to
one petal or sepal only, and this results in the
flower having an irregular shape. Gather a violet,
examine and compare the petals ; four of them
will be found to be nearly alike, but the lower petal
is much larger because it has grown into a tube
(called a spur) to secrete honey, and I need hardly
say that the honey is intended to attract the bees

so that the flower may be enabled to produce fertile seed. The enlargement of the lower petal gives the flower an irregular shape, and the same thing happens in the monkshood and many other flowers, where both the petals and sepals are thrown out of shape to form nectaries. In the orchid family this influence may be traced to a wonderful degree. The contrivances for insuring the fertilisation of their flowers are so many and various that books of the greatest interest have been written on that subject alone.

In the flowers we have hitherto noticed, both stamens and pistils are found, the petals are coloured, honey-glands exist, and some specimens also possess a powerful scent.

Such flowers are obviously very attractive to insects, and on that account they are called by modern botanists, entomophilous, which long word means that they are beloved by insects.

In sharp contrast to these gay and conspicuous flowers we may observe the very simple catkins of the birch, *Betula alba*. If we examine a twig of this tree in spring, we shall find two very distinct kinds of flowers (or catkins, as tree-blossoms ought properly to be called), one a stiff green spike standing upright, and the other longer and of yellowish colour, always to be found hanging down.

The former consists of a number of scales arranged on a central stem, and in the axil of each scale is the little pistil, with its pointed and divided stigmas. This catkin, later on, becomes the fruit of the tree, and sheds out with every passing breeze

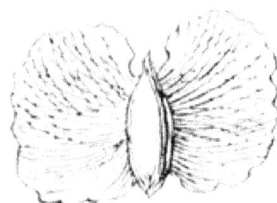

Natural Size.   Magnified.
BIRCH FRUIT.

its little winged fruits, which are carried far and wide and often sow themselves in rocky crevices, and appear able to grow and flourish with only a modicum of soil.

The pendulous catkin is very soft and loose, and on the inner surface of its scales we find the stamens, which in due time will shed from their anthers the fertilising pollen. Here then we see flowers which are not so attractive to insects, flowers in which the stamens and pistils are separated and developed in different catkins, and such flowers are termed monœcious, from *monos*, single, and *oikos*, a house.

The most interesting feature of these tree-blossoms is their fertilisation by the wind; the slightest puff of air liberates little clouds of pollen from the loose swinging anthers; these pollen grains become entangled in the upright catkins bearing the pistils, and the future seed thus becomes fruitful. There are many trees and plants which are thus fertilised by the agency of the wind, and they are termed by botanists *anemophilous*, from the Greek words *anemos*, wind, and *philos*, beloved by.

In the common bryony of the hedges, we get another example of a green inconspicuous flower. Gather a few sprays of this in early summer, taking care to keep the specimens of each plant separate. Take up one specimen and you will find each flower has a small green calyx, a minute corolla, and five little stamens; not one pistil can we find on the spray.

The flowers on the next spray look very

similar, but in them there are no stamens, the
centre of each flower being occupied by a small
pistil, and thus we learn that there are two distinct
sexes in the bryony plant, the one bearing only
staminate flowers, and the other
producing those bearing only
pistils. Such plants are termed
diœcious, from *di*, two, and *oikos*,
a house.

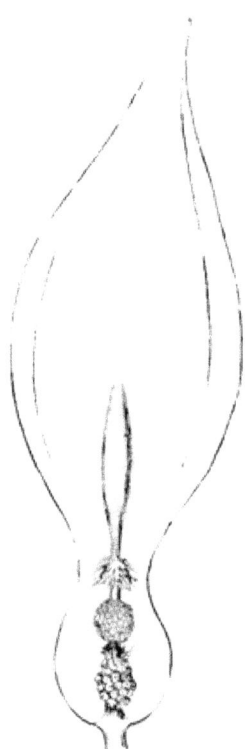

One of the earliest spring
flowers is the arum of the hedges,
known to village children as
"lords and ladies." Accustomed
as we are to bright-hued flowers
in our gardens and fields, it is
somewhat difficult to recognise
that the pale-green sheath of
the arum is a flower at all. It
consists of a beautifully-folded
spathe or bract, curving over at
the top, and if we remove that we find a central
stalk bearing a number of little naked flowers,
arranged in the order shown in the plate.

First, below the club-like apex, a few hairs tend-

WILD ARUM.

ing downwards, then the anthers containing pollen, and below these the pistils with protruding stigmas. The whole stalk is termed a spadix.

The outer green spathe forms a kind of prison, into which flies are enticed by the somewhat fetid odour which is exhaled by the flower. The flies easily creep in past the circle of hairs, which, as they point downwards, do not prevent their entrance, but, once in, these hairs are like a *chevaux-de-frise,* and hinder the escape of the insects. The flies in all probability carry upon their wings pollen from some other arum flower, and in their efforts to escape they brush off this pollen upon the stigmas, which thus become fertilised. When this has taken place the stigmas throw out a sweet juice upon which the insects feed; the anthers now shed out their pollen, with which the flies become covered; the hairs meanwhile have withered, and thus the flies, having done their appointed work in fertilising the flower, are free to crawl out and perform the same office for some neighbouring plant.

We have not space to do more than allude to certain plants, whose flowers never open and are self-fertilised. The common violet, for instance produces, in addition to its well-known fragrant flowers, certain inconspicuous blossoms, hidden under the leaves and close to the root, very seldom noticed by any but botanists, and known to them as cleistogamic flowers; these are fertile, and always produce seed. Other such plants are the woodsorrel and sundews.

It is interesting to observe the various ways in which flowers are protected from browsing animals, snails, and caterpillars by thorns, spines, prickles, and spiny bracts. The teasel secretes water in the bracts around its stem, which prevents ants from ascending to the flowers, and in many plants we may see quantities of small insects caught by a sticky gum exuded from the leaves and twigs.

Many delicate plants entirely alter the position of their flowers in order to protect them from rain. On a sunny day the wood-anemone holds its little snowy cup so as to receive the full sunlight, but on a damp day every blossom is closed and held downwards. We may observe this in the poppy, the blue-anemone, and nearly all composite flowers.

These are merely hints scattered over a wide

field of study, which some readers may like to follow out.

Objects to collect and examine:—Buttercup flowers, seed-vessel of wild-geranium, stonecrop growing on walls, flowers of bulbous plants, flowers of poinsettia, bougainvillia, salvia hormineum, arum. Examine various chrysanthemum flowers, sunflowers, asters and woodsorrel. Difference between hydrangea and poppy flowers, winter cherry (physalis); prepare flower dissections. Examine various cruciferous flowers and pea-shaped flowers, regular and irregular flowers, birch catkins, wild arum flowers, cleistogamous flowers, protection of flowers, position of flowers.

CHAPTER VII

POLLINATION

" When summer shines,
The bee transports the fertilising meal
From flower to flower, and even the breathing air
Wafts the rich prize to its appointed use."

COWPER.

POLLINATION

WE now come to the consideration of the real function of the flower of a plant. In whatever form it is developed, whether as a gay and fragrant blossom, in a dull foul-smelling structure like the arum, or as a green inconspicuous little floret like the grass, its main office is to reproduce itself by the formation of seed. We will first glance at some of the wonderful agencies that actively help in this work.

There are at least three distinct processes necessary for the complete formation of a perfect seed, and we must, I fear, persuade ourselves to learn some of the long words by which botanists speak of these processes. They are known as

pollination, fertilisation, and the growth of the
ovule. There is so much to be said about the
first subject, that I must leave the two latter for
a succeeding chapter.

Before seed can be formed it is necessary that
the powder contained in the anthers, which is
called pollen, should be transferred from those
anthers to the stigma or upper part of the pistil,
and this transference is called pollination. If we
examine a tulip or, better still, a buttercup, we
find the anthers and stigmas so near together
that the transfer of the dust-like pollen to the
sticky-looking stigmas can easily take place.
This would be called an instance of self-
pollination, but although cases of this kind do
occur in nature, they are not at all common.
As a rule, in order to ensure what is called
cross-pollination, the transfer of the pollen of one
flower to the stigma of another, many wonderful
and interesting arrangements exist even in some
of our commonest flowers.

Cross-pollination must be the case in such plants
as dog's mercury, because we find in a colony of
these plants—so frequently seen by the roadside—

that some plants have flowers with stamens only,
and others containing only pistils. Again, in the
hazel we may see how impossible it is for self-
pollination to take place, as, if we examine the
pistils, we find that they consist of scales bearing
stamens and pollen only, whilst somewhere close
by, on the same stem, hangs the pretty little red
flower which possesses the pistil and forked stigma.

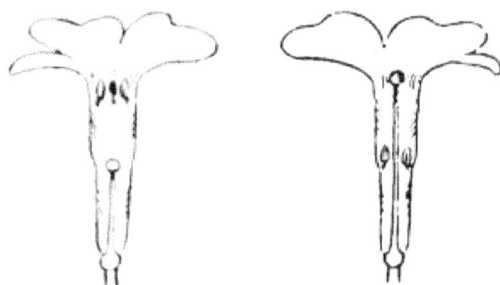

PRIMROSE.

If seed is to be formed in either of these flowers
and in many others similarly arranged, then the
pollen of one flower must be transferred to the
stigma of the other.

There are interesting facts to be learned about
the common primrose. When we examine a little
bunch of these flowers we find quite half of them
are what children call pin-eyed, meaning that the

stigma, which is at the end of a long pistil, is like
the head of a pin in the throat of the primrose.

Looking at the sketch, we see at once that self-
pollination is hindered by the fact that the anthers
in this flower being at the bottom of the tube, the
pollen they contain must be transferred by some
direct agency before it can come in contact with
any stigma.   Now let us examine the other flowers
in our primrose nosegay; we find the stamens in
these are placed in the mouth of the tube, and the
pistil is quite short and low down in position.   At
first sight it appears as if the pollen would fall
directly upon the pistil, since the stamens are
above that organ, but this is not exactly what
happens; the pollen of this particular form of
flower is shed before the stigma is mature, so
that when it has reached maturity the pollen
is all gone.

The arrangement of nature is as follows.   An
insect attracted by the sweet-smelling bank of
primroses will visit the flowers, thrusting its
proboscis down a pin-eyed flower until in so
doing its head has been dusted with the pollen
of the stamens; then withdrawing from that

flower the insect visits another near by, possibly
one with a short pistil; the pollen on its head
is now rubbed off and falls upon the stigma below
and pollinates it, for that is the term used when
this act takes place.

The pretty maiden-pink will help us still more

MAIDEN PINK.

clearly to understand how cross-pollination is pro-
moted in flowers containing both stamens and
pistils. Select a flower that has just opened, the
petals of which are spreading and fringed, whilst
from the centre of the flower a cluster of stamens
projects with the pollen mature and easily shaken

out of the anther lobes; the pistil is concealed in
the long tube, and in this stage there is no sign
of stigma. In a short time, however, if we examine
the flower again, we shall find the stamens have
shrivelled up, and in their place a forked stigma
appears, as shown in the sketch. Here again it is
obvious that the fact of the stamens ripening first
and expending their energy before the pistil is ripe
must mean, that in order to secure seed the pollen
from some younger flower must be transferred,
probably also by insect agency. It will give fresh
interest to our garden rambles if we remember that
the bees and flies we see hovering over the flowers
are not only collecting honey or feasting upon it,
but are also performing a very important office for
the benefit of the plants they are visiting.

We may now proceed to notice the various
agencies for the conveyance of pollen between
flowers.

These agencies are water, wind, insects and birds.

In an earlier chapter I gave an account of the
Vallisneria spiralis, which will serve as a type of
a water-pollinated flower.

Those pollinated by wind are, as I have said in

a previous chapter, called anemophilous (*anemos*, wind, and *philos*, loving). They are usually of small size and inconspicuous character, with very little or no scent, and devoid of colour; these are characteristics that are not always associated in the same species; thus in the hazel, which is a wind-pollinated flower, we find a bright yellow catkin (so well known to children as lambs' tails) and a small but bright red pistil.

Let us notice, however, how wonderfully these plants are adapted for this method of pollination; the stamens are usually hanging, and the pollen, produced in great quantities, is easily set free by the slightest breath of wind. The stigma of the hazel, of different grasses and of sedges are both forked and plumed, so that pollen grains floating in the air are readily intercepted.

The firs and pines are excellent examples of wind-pollinated trees. I remember once possessing a ripe male cone of the *Araucaria imbricata*, and ascertaining that it contained as much as a wine-glassful of pollen. Speaking about this fact to the gardener at the Pinetum at Dropmore, I was shown how this fertilising dust from the great

Araucaria (which was planted there in 1830) was carried by the wind for an amazing distance to a female tree on the other side of the garden, pollinating its cones so that they produced fertile seeds. In some of the Canadian pine forests, the trees shed forth such quantities of pollen in the flowering season that the ground becomes perfectly yellow. The early settlers, being unable to account for the strange phenomenon in any other way, ascribed it to showers of sulphur descending from the clouds. Even in our own country the foliage and undergrowth in the neighbourhood of fir woods is often thickly coated with the yellow dust falling from the male catkins of the trees; the structure of the pollen grains is such that they float very buoyantly, each grain being provided with two air bladders. I may mention in passing that this apparently wasted pollen affords a rich feast to endless species of bees and flies, and is in many cases stored up by them as food for their young grubs. The various adaptations for wind pollination will perhaps be better understood if we glance at the attractions which flowers offer to birds and insects.

Colour serves to render flowers attractive to insects, and to make them conspicuous; the bracts, petals, and sepals of flowers are usually of some light or dark colour quite distinct from the green tone of the foliage.

It has been ascertained also that plants which are pollinated by night-flying moths generally have white or light-yellow flowers so as to be easily seen in twilight.

One of the most interesting of these night-pollinated flowers is *Silene nutans*, the Nottingham catchfly. In the daytime the five narrow petals are curled up and look dead and withered, but as night comes on they change their position, and the flower has the expanded shape of an alpine pink. In this open condition it is visited by the moths which, flying from one flower to another, transfer the pollen, and thus accomplish at night what more frequently occurs in the sunlight; at daybreak the petals roll up once more, and one would again suppose the flower to be dead; but no, it will continue to open at nightfall until some moth finally succeeds in pollinating its blossom. A small species of moth¹ visits this catchfly in order to deposit its

¹ *Dianthœcia albimacula.*

eggs; these, by means of a very long ovipositor, it
places in the ovary, and in that somewhat inflated

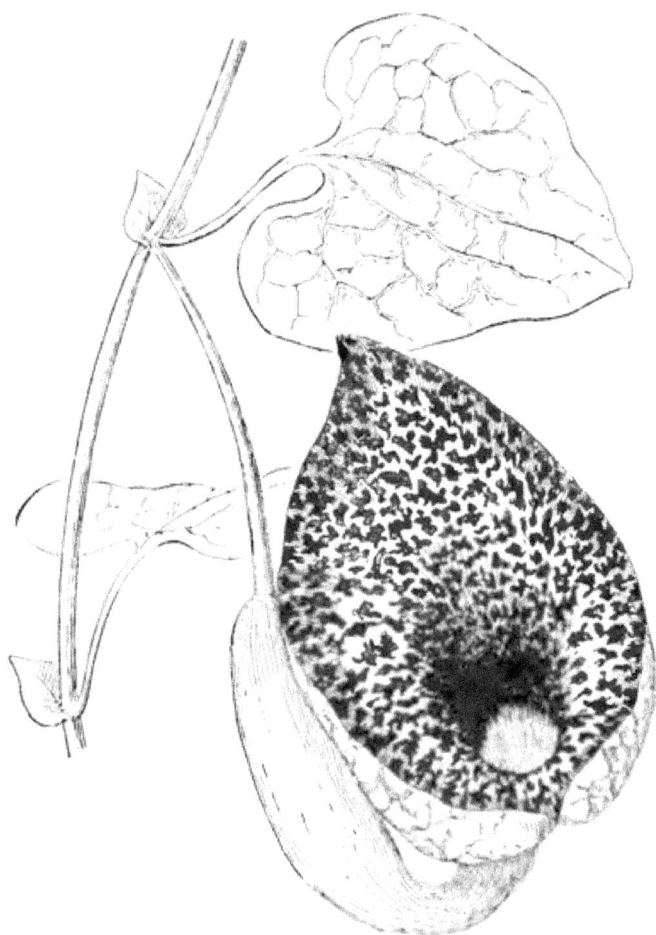

ARISTOLOCHIA.

cavity they produce microscopic caterpillars which
find shelter and nutriment in the strange nest.

When the caterpillars arrive at maturity they escape by biting a hole in the wall of the capsule, and creeping out, they seek for a suitable place in which to turn to chrysalides.

Scentless flowers usually have some equivalent form of attraction, such as honey, brilliant colour pollen in abundance, or the grouping of a number of small florets, in order to secure a conspicuous effect as in the ox-eye daisy, or hedge parsley.

Strong and varied odours are great helps to ensure pollination by insects. The bee-tribe and moths and butterflies are specially attracted by the sweet scents of roses, violets, carnations, and sweet-peas, and the powerful odour emitted by such flowers as the evening primrose, tobacco, and night-flowering rocket as evening comes on tends to guide the nocturnal moths to these and similar flowers. An odour may, of course, be pleasant to an insect which to us would be simply intolerable. The arum of the hedges, and those curious plants, the aristolochias and stapelias, all emit scents of the most fœtid description, as we think, but flies, on the contrary, are attracted by thousands, and hold apparently joyous revels in the blossoms

which they are pollinating by their frequent
visits.

STAPELIA.

A little care and patience in watching the visits
of insects to different flowers will soon be rewarded

by a perception of the tastes and likings of insect
life, and we shall gradually learn to expect to see
certain insects on the flowers they specially frequent.

I would call attention to the interesting fact that
if one agency fails to effect pollination, another is
adopted in order to attain the desired end. Thus,
when the flowers of the
common bartsia first open,
they are visited by insects;
but, in the later stages of
flowering, the pollen is blown
out by the wind, and the
neighbouring stigmas thus
become pollinated. We see
in the arrangement of the
flower of the St. John's wort
(*Hypericum*) a perfect type
of this provision against any

HYPERICUM.

possible failure of pollination. The stigma is surrounded by groups of stamens of unequal length;
those in the centre nearest to the stigma are as long
as the style itself, whilst those on the outside are
short, and these shed their pollen first, whilst those
in close contact with the stigma shed their contents

last. Thus we find that if insects fail to effect cross-pollination by means of the short and early opened stamens, it is secured by means of the longer stamens whose anthers are in close contact with the stigma. Again, when we stand under a sycamore tree, we may see that the green tassel-like flowers are having their pollen dispersed both by wind and bees.

We cannot draw hard-and-fast lines in nature, for although a special end may be kept in view, the various means and adaptations by which it is attained are a continual source of admiration and wonder to the reverent student of nature.

We have already seen that there are all kinds of devices by which the pollen of one flower may be made sure to reach the stigma of another; but, if by any means this crossing fails, if the weather is such that insects are scarce, or other conditions cause failure, then, in the case of many flowers, most curious contrivances are provided to secure seed by self-pollination. Truly this is one of the most beautiful of God's wonders in floral construction. One of the gems of my own flower garden is a lovely little Japanese toad-lily (*Tricyrtis hirta*).

In this flower there are three styles which stand
well above the stamens; the points of the styles
are bent over as in the plate, and the stigmatic surface grows mature before the anthers shed their
pollen: if, however, no insect visits the flowers, pollination is effected in the following way. The
styles bend down and place their forked points in
direct contact with the open anther-lobes (as shown
in drawing), the style assuming
almost the form of a semi-
circle. This is done very de-
liberately, for it is often fully
a week before the act is complete.

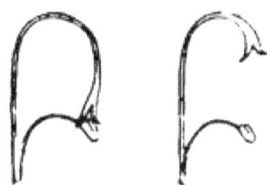

TOAD-LILY.
Stigma and Stamen.

Pollination is effected in tropical countries not
only by insects of many kinds, but by the lovely
tribes of humming-birds which abound in those
regions. Their slender, curved beaks are specially
adapted to penetrate the honey-laden flowers with
long-tubed blossoms, which could only be pollinated by some such agency.

Those who are within reach of the Natural History Museum at South Kensington may there see a
gallery filled with exquisite specimens of humming-

birds, arranged in cases, and some of the birds are shown as they appear in life, hovering over tropical flowers, drawing honey from their hanging blossoms, and performing the useful office of transferring the pollen from one flower to another, thus ensuring the fertilisation of the seed.

I might go on multiplying examples of the various methods by which seed is rendered fertile, but perhaps enough has been said to show what hidden force exists in flowers to enable them to attain the end for which they mainly exist, namely, the perpetuation of their species by means of seed.

Specimens to be obtained and compared with the descriptions in this chapter :   Buttercup flower, dog's mercury, hazel catkins, primrose flowers, male blossoms of pine trees in June, Nottingham catchfly, ox-eye daisy, bartsia, St. John's wort flowers, and Japanese toad-lily.

CHAPTER VIII

FERTILISATION

> "The men
> Whom Nature's works can charm, with God Himself
> Hold converse."
>
> AKENSIDE.

CHAPTER VIII

FERTILISATION

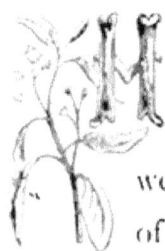

HAVING now considered some of the many wonderful arrangements by which the pollen of plants is dispersed, we will endeavour by tracing the course of the pollen-grains after they reach the stigma, to learn what is meant by the term " fertilisation of the ovules." These are the minute specks contained in the ovary which are to become seeds, and by means of which the plant will eventually reproduce itself.

To the naked eye the yellow pollen we see on the anthers of flowers appears as small grains ; but, when magnified, these grains are seen to be singularly beautiful, each little sphere having on its surface a chequered network and delicately sculptured patterns.

The forms, too, are as varied as the ornamentation.

Some plants have triangular grains, some oval-shaped and others many-sided.

I have given a few examples, and would

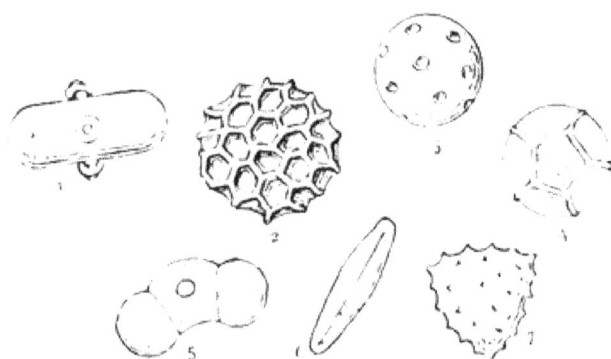

POLLEN-GRAINS.
1 Meurra.   2 Cobea.   3 Convolvulus.   4 Dianthus.   5 Pinus.
6 Althaea.   7 Buphthalmum.

specially call attention to the pollen-grains of the Pinus tribe (fir-trees), to which I alluded in the last chapter. These are remarkably buoyant, owing to the two little bladders with which they are furnished.

Now we are going to watch this yellow dust performing its appointed office in the central organ of a flower. In order to do so we will take

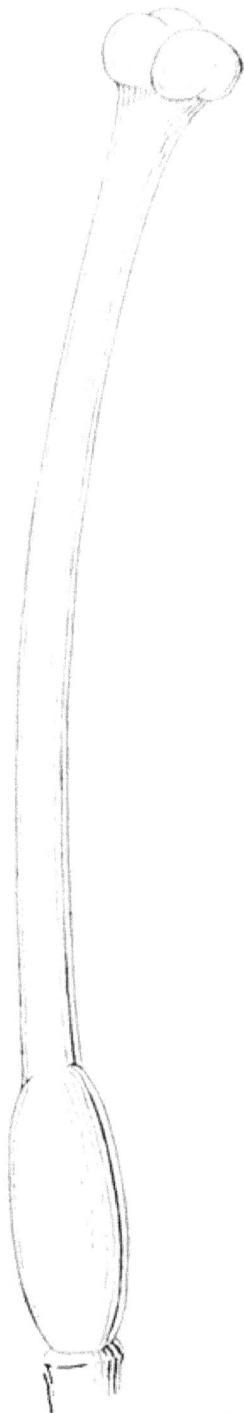

WHITE-LILY PISTIL.   SECTION OF PISTIL.

a white garden lily, and remove the petals, sepals,
and stamens, leaving only the pistil, which, as
shown in the drawing, consists of three parts, the
club-like stigma, a very long style, and its base the
ovary, which contains three cavities. In these
last we see a number of small, colourless spore-like
bodies termed ovules (from *ovum*, an egg), each
consisting of an outer coat, and a mass of cells in

the centre called the *nucellus*.

POLLEN TUBE.

An opening exists at one end
of each ovule called the micropyle
(meaning a little gate or entrance),
and this opening leads down into
the middle of the nucellus, where
lies what we may call the life-prin-
ciple, but what is known in botany
as the embryo-sac.

We need the aid of a microscope to enable us
to see how the pollen exerts its influence upon the
ovules.

If we place a drop of very weak sugar and
water upon a slip of grass, and sprinkle over it
some pollen grains of the common white lily,
then allowing the slide to remain for a few

hours in a dark place, it will be fit for our purpose.

When placed in the microscope we shall observe that many of the grains will have thrown out long thread-like tubes, and this is just what happens when pollen falls upon the viscid stigma of the lily. Referring to the section of a lily pistil we see that a pollen grain has rested on the stigma, and, excited into growth by the sweetish fluid which holds it there, it sends down a slender tube through the centre of the pistil, which is lined with a very delicate loose tissue of cells filled with starch, oils and food-materials. The pollen-tube is stimulated and fed by this nourishment stored up in the conducting tissue, and on it goes until, passing through the micropyle, it enters the embryo-sac of one of the ovules, adheres to it, and renders it fertile.

Only one grain is shown in the drawing for the sake of clearness, but of course each ovule is sought out and fertilised by a pollen-tube. With infinite variation this process takes place in every flower, so that even the commonest weed affords evidence of the marvellous provisions made by

an All-Wise Creator for the preservation of
species.

The time occupied by the passage of the pollen-
tube varies considerably. In the fir tribe it takes
nearly twelve months, in the hazel-nut and orchis
it requires several weeks, whilst in many other
plants the whole process is completed in a few
hours.

One of the first results of fertilisation is a rapid
withering of the style and flower ; the great end
of the flowering period has been attained, and so
without further expense of energy the bright petals
die away.

At the same time other external changes take
place, which are obvious to every observer of
nature. The lower end of the pistil, known as
the ovary, begins its second growth, and in a short
time swells into a large structure, the shape of
which varies much in different species of plants.
Finally, the ovary changes colour and develops
other characteristics quite different from its
former conditions. These characters have reference to the distribution of its seeds, and in our
chapter on fruits we shall learn something about

the interesting botanical significance of the various
hard and soft fruits, and see how they all arise
from fertilisation.

Take, for example, the flower of an apple
immediately after fertilisation is effected. The
petals fall off, the styles shrivel up and the ovary
rapidly enlarges; the tube of the calyx becomes
fleshy, and finally the well-formed apple is
produced. The change, however, does not end
here; in this stage of development the little apple
is bitter and is charged with a vegetable acid. As
the fruit grows on, however, this acid changes into
sweet juice varying in flavour according to the
species of apple.

Now let us examine the interior of the ovary
and see what changes have arisen as a consequence
of fertilisation.

The egg cell which has received the pollen
grain becomes filled with an embryo, whilst the
thin delicate coat of the ovule develops into strong
seed-coats.

The embryo is the first germ of the young plant
that is to be. It is a tiny speck indeed in its
beginning, but deeply interesting to us when we

realise that, because it possesses life, it will grow
on and on, and result, according to its species,
either in a plant but a few inches in height, or
in a grand forest-tree which may give shelter
to man and animals for hundreds of years.

The naked eye can scarcely trace any indications of form in the embryo, but when dissected
and examined with a lens it is seen to consist of a
tiny plant, root, stem and leaves (cotyledons).

The size of the embryo in comparison with the
other part of the seed is a point which should be
observed.

As the embryo develops it absorbs the special
nutrient or reserve tissue that exists in all ovules ;
a bean embryo, for example, rapidly absorbs all
the nucellus of the ovule, so that at length the
seed-coats contains nothing but the embryo, the
two cotyledons of which are thick and filled with
stores of food for the first growth of the seed.

I would advise students to plant a few broad
beans in a little damp cocoa fibre, and carefully
watch their growth. It is advisable to dissect
these beans successively at different stages, so as
to watch the development of the radicle (root) and

plumule (young leaf-bud). Place the seed in what
position we may, the radicle will always find its
way down into the earth, while the plumule obeys
its vegetable instinct, and rises into the air. The
embryo of the castor-oil bean and that of the
cocoa-nut do not, however, use up all the nutritive
matter in the ovule as the broad bean does, so

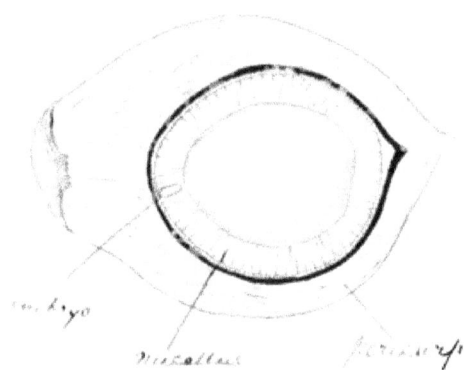

SECTION OF COCOA-NUT.

that when the seed is ripe we find inside it, not
only the embryo, but also a quantity of cheesy
matter known as *albumen*, and seeds of this kind
are hence called *albuminous*, whilst peas, beans
and hazel-nuts are classed as *ex-albuminous* (without albumen).

An interesting development consequent upon

fertilisation is a growth which occurs in some
plants from the base of the ovule. The pretty
red coverings of the seeds of the spindle-tree, and
the bright berry-like structure on the seeds of
the yew-tree are examples of this growth, which
is known botanically as an aril (from *arillus*, a
wrapper). In the willows this aril is a very

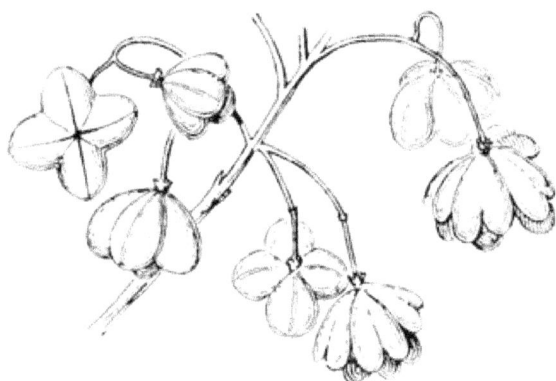

SPINDLE-TREE.

lovely covering of silky hairs, these serve to float
the seeds on the atmosphere at every puff of
wind.

The pretty lace-like covering on the nutmeg is
another example of an aril, better known to us in
the form of the fragrant spice called mace.

The style, which in most plants dies as soon as
the ovules are fertilised, is in other cases persistent,

as in the hedge-climber called travellers' joy.
The white, feathery - looking seeds owe their
special character to the persisting styles, which,

NUTMEG AND MACE.

after fertilisation, grow into the bunches of fluffy
seeds, which hang in profusion on hedges in the
country.

I will conclude this chapter with a reference to

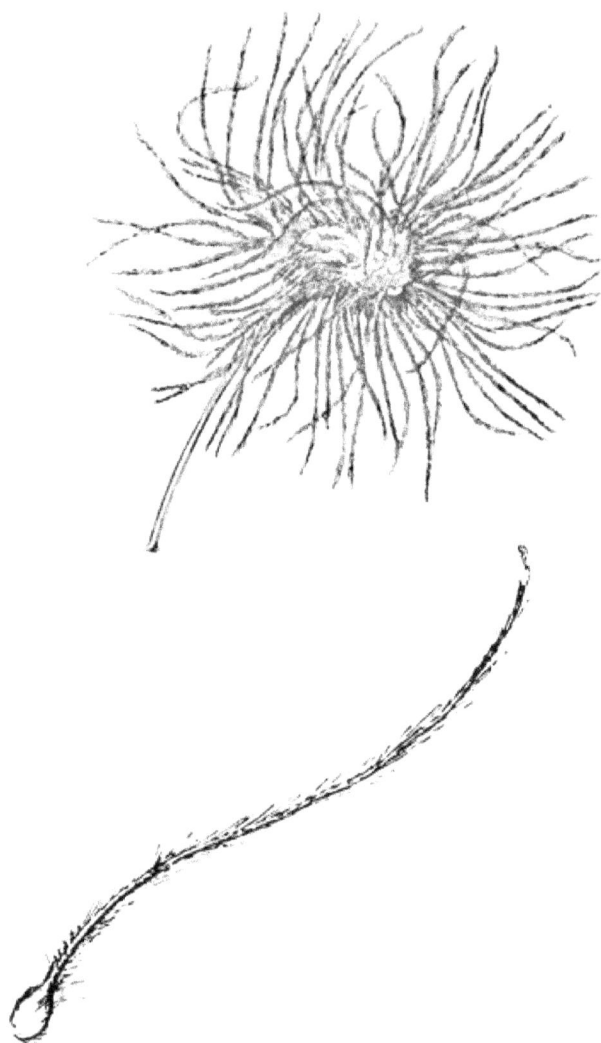

CLEMATIS OR TRAVELLERS' JOY.

a change of quite a different character. Not unfrequently, fertilisation results in the suppression of certain chambers in the ovary, and in the consequent failure of the development of the ovules.

A cross-section of a young oak ovary shows a three-chambered structure, each cavity containing two ovules, so that the ovary in this stage contains six ovules in three chambers. Soon after the act of fertilisation, one of the fertilised ovules takes the lead in growth, starves the other five ovules, and, as it grows, pushes the partitions of the other chambers aside, and gradually fills up the whole interior, converting it into a one-celled structure. This change happens also in the birch; its two-chambered ovary becomes one; and in the lime, though at first it has a many-chambered ovary, yet in the ripened fruit there is rarely more than one to be found.

In a few plants, changes of quite an opposite character take place. In the ovary of the datura,[1] for instance, we find two cells; after fertilisation, two false or spurious partitions are developed, dividing the original two-celled structure into four

[1] Thorn-apple.

parts, and as a consequence we get a four-chambered fruit. The same change takes place in some of the pea family.

Specimens to be observed:—Examine pollen grains, with lens or microscope, dissect white lily, flower-pollen on glass slide. Observe changes in growing apple, plant broad beans, castor-oil seeds, and maize; examine spindle-tree berries (*euonymus*), yew-tree berries, willow seeds, nut-meg, and mace; travellers' joy (clematis), section of oak ovary in the pistillate flower. Examine birch catkins and lime-tree flowers. Datura seed-vessel.

CHAPTER IX

FRUIT

" Here, as I steal along the sunny wall,
   Where Autumn basks, with fruit empurpled deep,
   My pleasing theme continual prompts my thought ;
   Presents the downy peach : the shining plum ;
   The ruddy, fragrant nectarine : and dark,
   Beneath his ample leaf, the luscious fig,
   The vine, too, here her curling tendrils shoots,
   Hangs out her clusters, glowing to the south,
   And scarcely wishes for a warmer sky."

JAMES THOMSON.

FRUIT

F we are shown a collection of delicious apples, pears, grapes, peaches and cherries, we form a very appreciative opinion of the use and function of fruit, but that opinion is somewhat modified when we are shown a basket of poppy-heads, acorns, the light downy seeds of the thistle, the small dry carpels of the buttercup or the winged fruits of the maple. We usually connect the term fruit with some luscious product of the vinery or kitchen-garden, and we may include as such the brightly-coloured berries of the hawthorn and wild rose, which are so conspicuous on trees and hedges in autumn; but if we examine the subject botanically we shall have to widen our ordinary conception of the term.

There is probably no part of a plant so difficult to understand as its fruit, and this difficulty is due to those many changes which I described in my last chapter. A very general definition of fruit is that it consists of the ripened ovary, and this will be found to be correct in a great number of cases, but this term is not exactly wide enough to express the general formation of all fruit. In some cases it is composed of the ripened ovary with the parts of the stalk or the original flower, enlarged or incorporated in the structure of the fruit, but in other specimens we find the ovary, although present, very little enlarged, and playing but a minor part in the ultimate character of the mature fruit.

No fact seems so emphatic to the observant botanist as that which upsets his artificial rules and classifications of plants and the parts of plants. We say, for instance, that fruit is the ripened ovary, and yet directly we leave our books and go out to study botany in the fields and woods, we find a large group of fruits perfectly innocent of any such structure. The firs and pines have no organ of this kind, and yet their fruits are most important and extremely interesting. Scarcely any part of a plant varies so much in different species as the fruit does. Although leaves may be found of every size and shape, they still have some general similarity of form, but we hasten to observe what an immense contrast there is between the huge *Musa* fruit (banana) and that of the oak (acorn), although the former is, compared to the latter, but a poor weakly plant.

Again, let us note the difference between the cocoa-nut palm fruit, a nut, which with its outer husk is almost as large as a peck measure, and that of the St. John's wort or any other of our native wild flowers.

These differences in size have their counterparts in other directions. We generally think of fruits as being soft, luscious, and pleasant to the taste. Many fruits of delightful colour and texture are, however, bitter as gall, and possess highly noxious qualities. I well remember gathering a plateful of rich purple berries from a plant I discovered in one of my childish rambles and carrying them

home as a great prize; I was not a little disap-
pointed when I learned that they were the
poisonous fruits of the deadly nightshade; their
deceitful resemblance to plums, as well as the
berries of the woody nightshade to red currants,
make these two of our most dangerous native
plants.

As offering very distinct contrasts to the above,
we may note the dry membranous fruits of many
of our forest-trees, the hard nuts of the hazel and
walnut and the leathery husk of the chestnut.
Again, the shape of fruits is wonderfully diversified. We have round and oval apples, plums, and
gooseberries; the linear seed-pods of the cabbage,
cauliflower, wallflower, peas and beans, and other
plants in endless varieties of forms.

There are contrasts again in the smooth surface
of some fruits and the hairy coats of others where
the roughness is due to hooks, prickles and other
contrivances. How different, too, is the airy
pappus of the dandelion to those heavy fruits
which drop like stones and are to be found lying
exactly beneath the branches where they have
ripened.

FRUIT 215

These differences in external form are multiplied
when we examine fruit more minutely. We shall
then find a useful dividing line in the manner in
which fruits allow their seeds to escape. In one
large division the **fruit when perfectly ripe splits
open and allows the seed to fall out; such fruits
are termed dehiscent** (from *dehisco*, I gape). In
the other division the fruit **remains closed, and the
substance of it must** decay **before the seeds can
escape and** grow ; **these are** classed **as indehiscent**
(I gape not). Before referring to a few examples
of each division we will endeavour to distinguish
clearly the various parts **of a fruit and learn their
proper botanical names.**

We must be careful not to confound the seed
and the coats of the ovary ; the latter is termed the
pericarp (*peri*, around, *karpos*, a fruit). In some
fruits this pericarp is developed into distinct coats,
or layers. In a peach, for instance, the outer coat
is rough and hairy, this is called the epicarp (1) (*epi*,
upon, *karpos*, a fruit) ; the middle coat is the
succulent delicious fruit, and is known as the mesocarp (2) (*mesos*, middle, *karpos*, fruit), whilst the inner
coat is the hard stone, or endocarp (3) (*endon*, within,

karpos, fruit), and inside it lies the kernel, or true
seed. As a type of a dehiscent fruit we may
select a pea-pod; here we get no division of the
coats into distinct parts, the pericarp is dry and
tough, and when perfectly ripe it bursts open,
and allows the seeds to escape.

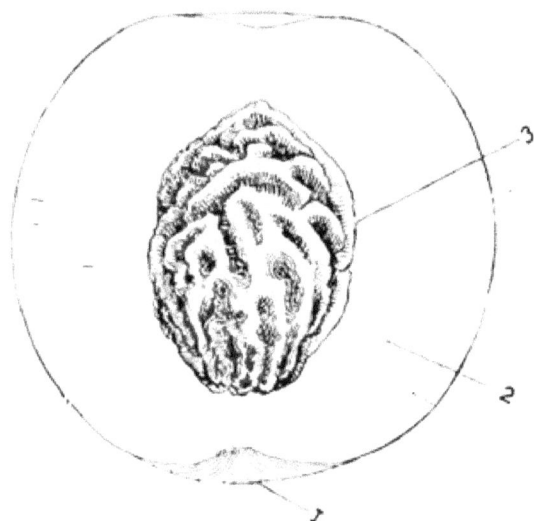

SECTION OF PEACH.

It would be very interesting to make a collection
of various seed-vessels, and note the immense
variety of ways in which the seeds find their way
out of the dry capsules. A poppy-head, campanula and antirrhinum sprays, henbane, colum-
bine, stramonium, and many other plants afford
good examples.

The woody pear is the hard fruit of a New
Holland plant which splits open to release the
seeds. The horse-chestnut is a conspicuous instance
of a dehiscing fruit, the rough prickly part is the
pericarp, and when the fruit is mature this splits
open and allows the two large chestnuts (seeds)
to escape. In the sweet-
chestnut we get an alto-
gether different structure.
If we pick up one of its spiny
burrs, we hold in our hand
what is called in botany an
involucre (from *involucrum*,
a cover), that is, a number
of bracts which have grown
together and formed an
outer case to the fruit. The
acorn-cup is an involucre, and

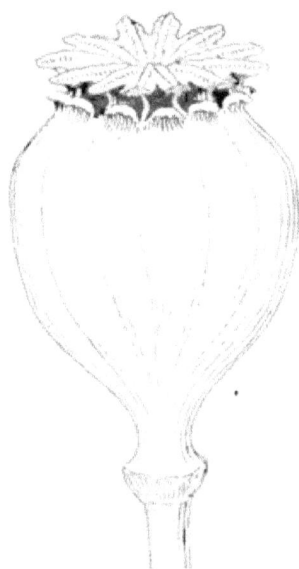

POPPY CAPSULE.

we may find other good examples in composite
flowers and those of the umbelliferæ. The small
green whorl in which a daisy-flower is set is, therefore, not a calyx, but an involucre consisting of
minute bracts grown together. The true fruit of
the sweet-chestnut is enclosed in a mass of spiny

bracts, and thus differs entirely from the pericarp of the horse-chestnut; if we wish to speak of it correctly we must call it either a cupule or involucre. We will now select a few examples of fruits that are indehiscent.

On the outside of an orange we find the yellow coat of the pericarp, next to it is the white mesocarp, and inside is the juicy endocarp,[1] in which the seeds are embedded. When an orange falls to the ground these different coats simply decay, and the seeds are aided in their efforts to grow by the succulent

WOODY PEAR.

flesh of the fruit, which affords them moisture and nutriment. The hazel-nut is a fruit of another texture altogether. The hard shell is the pericarp, and the one or two seeds within it must

---

[1] Strictly speaking, the endocarp of the orange is a thin membrane, and the pulp grows from it and fills up the ovary cavities.

remain enclosed there until the shell decays and the kernels can germinate and become new plants.

In the currant, gooseberry, and date we find examples of indehiscent fruits with a sweet fleshy pericarp. In the date there is only one seed in each fruit, and a curious thin endocarp can be observed enveloping the solitary seed. Many allied species, as well as the date, possess this sweet pericarp, which must decay in order to liberate the seeds, and in the case of succulent fruits the process is frequently assisted by the fruit-eating birds.

It may be well to draw attention to the very simple kind of fruit possessed by the buttercup and other similar plants. It is a dry membranous pericarp, and inside one seed exists free from the pericarp; this remains closed, like other forms of the indehiscent types, and technically this fruit is known as an achene (from *achanes*, not gaping), and it is well named, as it remains closed until decay enables the growing radicle to break through the pericarp and enter the ground. The curious after-development of the strawberry fruit is worth a little careful study.

This flower is known as apocarpous (*apo*, apart, *karpos*, fruit), consisting of a number of distinct ovaries each with one ovule ; these ovaries when ripe are exactly like the achenes of the buttercup, but they are developed upon a receptacle which, when fertilisation has taken place, begins to dilate and swell, with the result that the little achenes are gradually scattered over the surface of a large fleshy receptacle which, as it nears its time of perfection, becomes of a most tempting crimson colour. The little seed-like dots we notice on the strawberry are distinct and perfect fruits embedded in a sweet succulent floral receptacle. Thus we find that the strawberry, speaking botanically, is not a berry, but a collection of achenes, the term "berry" being usually restricted to such fruits as the currant and gooseberry. For this reason the strawberry and the common fig are sometimes termed spurious fruits, for in these the soft pulpy flesh is really the receptacle and the little round so-called seeds are the true fruit.

There is a very different formation in the pineapple, since this fruit is the development of an entire spike of flowers ; these in their early stage

are crowded together on the flower stalk, but as
time goes on they coalesce and fuse, with their
ovaries, bracts, and receptacles, into a succulent
mass, the various parts of which can be well defined
if we cut a section through a pineapple before it
is quite ripe.

This chapter may fittingly conclude with a brief
reference to the ultimate purpose of these varied
forms and textures of fruit, for that they each have
their special work, and that there is a meaning for
every form, is a truism we may accept without
doubt. The fruit is in reality the storehouse for
the seeds, the latter being the vital part of the
plant. If we review the life-history of a plant,
first its producing flowers, then the special and
intricate processes of pollination and fertilisation,
and subsequently the growth of that wonderful
little part, the ovule, into a seed, and further if we
reflect that the whole strength of the plant has
been concentrated on producing that seed, we
shall then comprehend the true significance of
fruit.

The seed is first stored up in the recesses of the
ovary; clearly then the ovary, which subsequently

becomes the fruit (pericarp), is intended to protect
the seeds, and it is interesting to note some of the
various ways in which this protection is afforded.
Take first the soft and sweet fruits so plentiful in
the autumn ; this edible sweet flesh is not developed until the seeds are quite ripe.   All through
the period of growth and ripening the pericarp
is hard or stringy or it may be also sour or acid.
This is especially true of hedgerow fruit, such as
crab-apples, sloes, and wild pears, texture and juice
alike affording complete protection.

Again, such fruit as the walnut and chestnut
are protected by their rough covering and hard
shells, and many others have their outer coats
covered with prickles and spines for the same
reason.   The most extreme case is perhaps that
of *Mucuna pruriens*, a leguminous climber found
in the tropics ; this has downy pods not unlike
those of a sweet-pea, and these pods are covered
with brownish hairs which, if incautiously touched,
enter the pores of the skin and cause a most
intolerable irritation ; a truly formidable protection this to the seed.

Let me now point out how the seed is protected

in some of the pine family firs , where there is
no pericarp. During the growth and development
of the pine seeds, the woody cone is rich in resin,
and should an enterprising nuthatch try to peck
out the seeds, he finds his beak covered with the
resin and his effort baffled.

Protection is also afforded to the seed by the

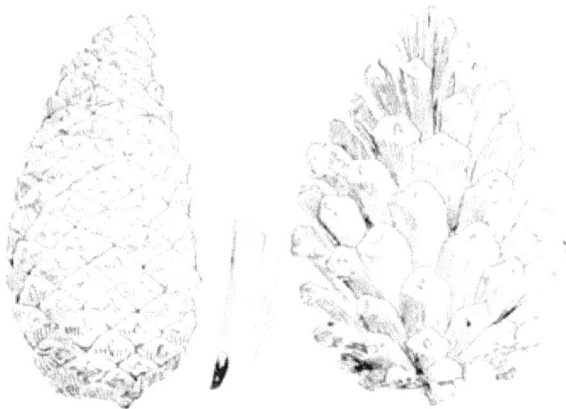

PINE CONES.

movements of fruit after fertilisation, and of this
the cyclamen flower affords a good illustration.
As soon as fertilisation has taken place the flower
stalk coils up like a watch-spring, and the seed-
pod is thus placed safely beneath the leaves to
ripen.

In crevices of old walls we may often find that

charming little wilding, the ivy-leaved toad-flax ; it has a highly intelligent method of protecting its seeds. When the flower is fertilised its stalk bends its point round to the wall, and places the tiny ovary in a cranny of the brickwork to mature and ripen its seeds. These are but two instances, out of hundreds, of plants whose fruits are protected by what we call instinctive movements.

It is of essential importance to young seedlings that they should have sufficient soil, light, and air, to ensure their healthy growth. To begin life directly under the leaves of the parent plant is to court failure and starvation, and so we find in the fruit that wonderful provisions are made to ensure the dispersion of the seed when it leaves the parent plant, and so endless are the contrivances for the dispersion of fruits and seeds, that it will be needful to devote the next chapter entirely to that subject.

Objects to collect and examine :—Compare various fruits, fir-cone, banana, acorn, seeds, and berries, &c. Examine a peach and pea-pod. Collect seed-vessels, horse-chestnut, sweet-chestnut,

daisy-flower, orange, hazel-nut, date-fruit, strawberry, pineapple.

Observe seed coverings, pine-cones, cyclamen
stems after flowering, seed capsules of ivy-leaved
toad-flax in wall crevices.

CHAPTER X

DISPERSION OF FRUITS AND SEEDS

" Who gave the thistle's feather'd seed its plumes,
   That wing-like waft it on each gentle breeze
   To sterile yet to it congenial soils,
   Investing them with purple beauty, rife
   With fragrant treasures for the wild bees' store?"

T. L. MERITT.

CHAPTER X

PURPOSE in this chapter to explain some of the many remarkable ways in which plants are enabled to scatter their fruits and seeds. The chief agencies which assist in carrying out this purpose are wind, animals, birds, running water, and moisture in the atmosphere. We shall find that many seeds are furnished with certain outgrowths and peculiarities which are specially adapted to the action of these agencies, with the result that such seeds are distributed far and wide. We will first examine some of those fruits which are scattered by animals; this end is generally attained by means of hooks and curved spines on the outside of the fruit.

229

Perhaps one of the most remarkable instances of
this class is the seed-pod (or capsule) of the Mar-
tynias. During the visit of the Prince of Wales to
India, a panther killed in one of the shooting
excursions was found to have quantities of long-

SEED-POD OF MARTYNIA.

hooked seeds attached to his skin: these must have
been brushed from a plant of *Martynia proboscidea*,
which has sharp curved horns three or four inches
long.

Another species called by the Italians *Testa di*

Quaglia, or quail's head, sows itself in a similar manner by clinging to moving objects.

Many common hedgerow plants have their fruits armed with quite formidable hooks, so that creeping or flying creatures may be made unwittingly the means of distributing the fruits. The burdock is a most persistent plant in this respect, each of its numerous fruits being covered with long hooks which successfully retain their hold of our clothing if we happen to brush past the plant when covered with its troublesome burrs. Other examples are the rough seeds of the forget-me-not, agrimony, enchanter's nightshade—a great pest in gardens—and all the bedstraw tribe.

These plants, we may observe, are low-growing and herbaceous, quite distinct in the matter of position from the tall trees and shrubs which depend upon the wind to scatter their seeds.

We are all familiar with the winged fruits of the sycamore; they are to be seen in early autumn. The clusters are first of a pale green, and then the seeds[1] often attain a flush of pale crimson which

---

[1] In botany the fruit of the sycamore, maple, ash, &c., is called a *samara*, and is properly speaking a winged *achene*.

adds much to the picturesque beauty of the tree. The equinoctial gales separate the seeds from their stalks, and away they go far and wide, borne up by the delicate membrane attached to the seed which catches the wind, and is carried by it to a great distance from the parent tree. In the same way the winged keys of the ash, being very light, are borne by the autumn gales to strange habitats, so that the tree may often be found growing on

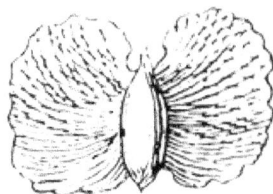

Natural Size.   Magnified.

BIRCH SEED.

church towers, in ruins, and on crags inaccessible to man.

The pinus tribe of trees have seeds with wings lightly twisted so that, if we hold up a dry fir-cone, the seeds descend from it with a whirling motion like small shuttlecocks.

The winds which blow strongly in mountainous places carry these seeds before them, and are thus ever renewing the pine-forests by sowing the pro-

ducts of their cones on bare tracts of land. The
lightest of all tree seeds is that of the birch; it is
gifted with two wings or membranes, so that it
floats in the air before the lightest breeze, and this
may account for the wide distribution of the tree
which has been found growing from Mount Etna
to Iceland and Greenland. I may give an instance
of a common which, twenty years ago, was covered
only by furze, broom, and brake-fern; about four-
teen years since, a shower of birch seed must have
been strewn over the ground, and now it has become a wood, shutting out the distant views and
quite altering the character of the landscape.

The wind again is the agency for the dispersion
of the seeds of such plants as the common groundsel; here it may not be uninteresting to note the
beautiful provision made in regard to the buoyancy
of the seeds. These winged structures which the
wind so lightly blows into the air must attain a
certain altitude from which they can be successfully launched, and therefore we find that a large
class of low-growing plants have their seeds furnished with accessories in the form of light silky
down or hairs.

Most of the plants known as *composite* have their seeds thus feathered, and amongst them are those plagues of the farmer, the thistle, dandelion, goat's-beard and others. The dandelion may serve as our example, and I would advise my readers to

PARACHUTE.

watch carefully the variations of position in the flowering stems. Whilst the flower is still fully expanded the stalk remains in an upright position so that it is conspicuous and likely to attract the notice of insect visitors. After the florets are fertilised it gradually lowers itself until it lies on the ground under the leaves for a period of ten or twelve days. During this time the seed-vessel matures and ripens, then the stalk rises to the erect position once more, and the beautiful downy globe expands into a soft fluffy

DANDELION SEED.

ball of seeds hanging so loosely that the first breeze carries them away, and their descent into the ground is curiously provided for. Persons have sometimes alighted on the earth from a balloon by means of a parachute, a machine which closely resembles an open umbrella with a car at the lower end. Now the dandelion seed descends in a similar manner, touching the ground first with its lower end, the weight of the seed enabling it to drop into some hole in the soil, and the spiny projections at the upper end preventing the feathery part of the

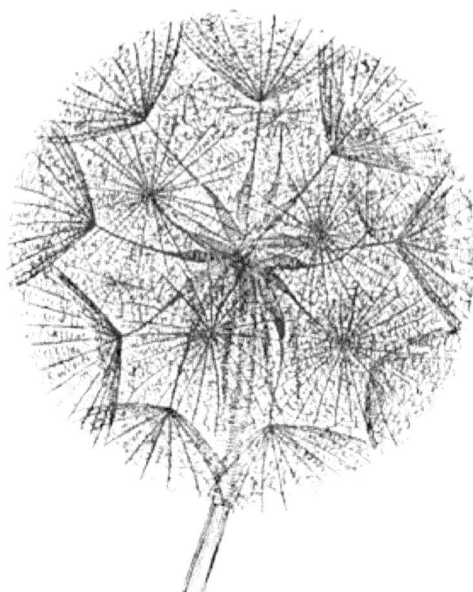

GOAT'S BEARD.

seed from dragging it out again. The common goat's-beard is perhaps the most beautiful English example of fruit with a downy pappus. A single flower will produce a sphere as large as a cricket-ball, and each seed is furnished with a starlike crown of branched feathers which the wind can bear away to a considerable distance.

The handsome willow-herb, which adds so much colour and beauty to our river banks, bears its seed in long, narrow pods, and these, when ripe, split up into five segments which, curling back as they open, leave the downy seeds free to be carried off by the passing breeze.

Bird agency in seed dispersion is a most interesting subject, and one can but admire the wonderful way in which the services of winged creatures are made available.

Succulent berries and sticky fruits are highly attractive to many kinds of birds, and whilst they revel upon the sweet, soft flesh of the berry, the seeds which they swallow with it are enabled to resist the action of digestion by a hard covering which protects the kernel until the shell shall decay and allow the seed to germinate. In this way I find my garden in early spring quite thickly strewn with the seeds of the Irish ivy, always a favourite food of the common wood-pigeon which is so frequently to be heard cooing in my woods.

The seeds of aquatic plants often cling to the

feathers of birds that visit pieces of inland water, and are widely distributed by them in their flight from one lake to another.

Darwin has shown by careful experiment that the mud clinging to the feet of various birds almost always contains seeds. A wounded partridge had

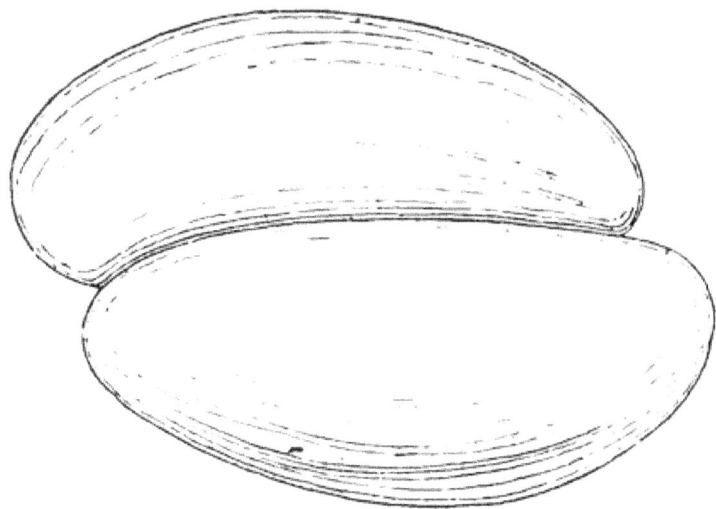

COCOS DE MER.

a ball of earth weighing six and a half ounces adhering to its legs. From this earth Darwin reared no less than eighty-two separate plants of five distinct species. Seas and rivers also do their part in dispersing seeds. The huge nuts of the *Cocos-de-mer* palm, which grows only upon the

Seychelles Islands, are often thrown upon very distant shores. This nut is said to take ten years to come to perfection; it is exceedingly hard, and sometimes weighs as much as forty pounds. The common cocoa-nut is often found growing on the shores of coral and other islands in the Pacific Ocean, and owes its position there to the buoyant nature of the nut, which floats un-injured in the sea until it finds a resting-place and a home on some atoll or island. In this way the cocoa-palm has spread to such an extent that it is now perhaps the only palm common to the western and eastern hemispheres. West Indian seeds and fruits have even been thrown upon the Nor-wegian coasts, and, but for the unsuitability of the climate, there is little doubt that tropical trees and plants might sometimes be found growing even so far north. It is obvious that the seeds of all vegetation on the banks of rivers, small running streams and lakes, must be liable to very wide distribution. Darwin made many interesting ex-periments as to the length of time seeds could retain their vitality when floating in fresh or salt water. Ripe hazel nuts germinated after being

ninety days in water. An asparagus plant with
mature berries, when dried, floated for eighty-five
days, and the seeds afterwards grew vigorously.
Out of ninety-four plants experimented upon,
eighteen floated for more than a month and some
for three months, their germinating power not
being destroyed. In quite a large number of
species the plants themselves possess the means
necessary to distribute the seed. It is true the
distance traversed by each seed may not be great,
but it is sufficient to give the seed a new field of
growth. This power varies in different species. It
is perhaps best defined as elastic force, and in the
majority of cases the seed is actually thrown away
from the parent plant by the expenditure of this
force. The seed-pod is generally in a state of
tension, due to the gradual drying up of the
tissues. Then a puff of wind, a slight blow, or
even a change in the atmospheric condition of the
air, gives the final impetus, causing the pod to
burst with such force that seeds are thrown out in
all directions. The fibro-vascular cords are often
found crossing the pod in an oblique direction, or
even in a spiral manner, so that finally, as they

shorten through dryness, they act upon the walls
of the legume and we see the result in such dried
pods as those of the sweet pea, broom, and
laburnum.

The pansy has a three-valved seed-pod, and as
it dries the edges of the valves press upon the
polished, hard-shelled seeds and they are squirted
out with a jerk to a distance of several feet. I
was once greatly puzzled by a strange, crackling
sound in my room, and after a few minutes' search
I discovered it was caused by a fusillade of pansy
seeds striking against the sides of a small box in
which I had placed the capsules to ripen. It is
worthy of notice that the capsule hangs down to
protect the seed-valves from rain; but when the
seeds are matured the capsule rises to an upright
position so that they may be projected far and
wide. A conspicuous example of the elastic force
of which I have spoken is seen in the British
balsam, *Impatiens Noli-me-tangere* (touch-me-not).
When its seeds are mature, the valves of the
capsule curl up in a spiral form with such force as
to project both themselves and the seeds through
the air many feet from the plant dropping the

16

seeds by the way. On a hot summer's day one
may hear the dispersion of seeds! The furze and
broom pods, the sweet peas, and especially fir tree
cones, make quite a loud report as they split and

BROOM AND SWEET-PEA PODS.

scatter their contents. The tension causing these
explosions is in some cases brought about by the
fluids inside the fruit. This is the case with the
squirting cucumber, which, when fully ripe, is so

distended with fluids that the slightest touch or
movement is sufficient to cause it to break away
from its stalk, and then the whole contents are
ejected with great force, so that the seed is thrown
some distance. The extent of dispersion is very
limited in those plants that are dependent upon
the varying moisture of the air. Such plants are
usually furnished with special awn-like [1] appendages ; these are hygroscopic [2] in their nature, and
the difference in the amount of moisture in the air
lengthens and contracts these apparently moving
organs. When the seeds fall from an ear of barley
they lie thickly strewn around the bottom of the
stem, and, were they to take root there, they must
inevitably choke each other ; but each awn is
thickly set with bristles, and as the morning sun
shortens and the evening dew lengthens the hair-
like awn, the prickles only allow the awn to move
in one direction, and the seed which is attached to
it is slowly but surely drawn many inches away.
What is popularly called the dancing oat is another
curious example of this hygrometric property. If
a dry seed (or oat) is placed for a moment in

[1] The beard of corn.　　　　[2] Sensitive to moisture.

water, and then laid on a smooth table, it will be

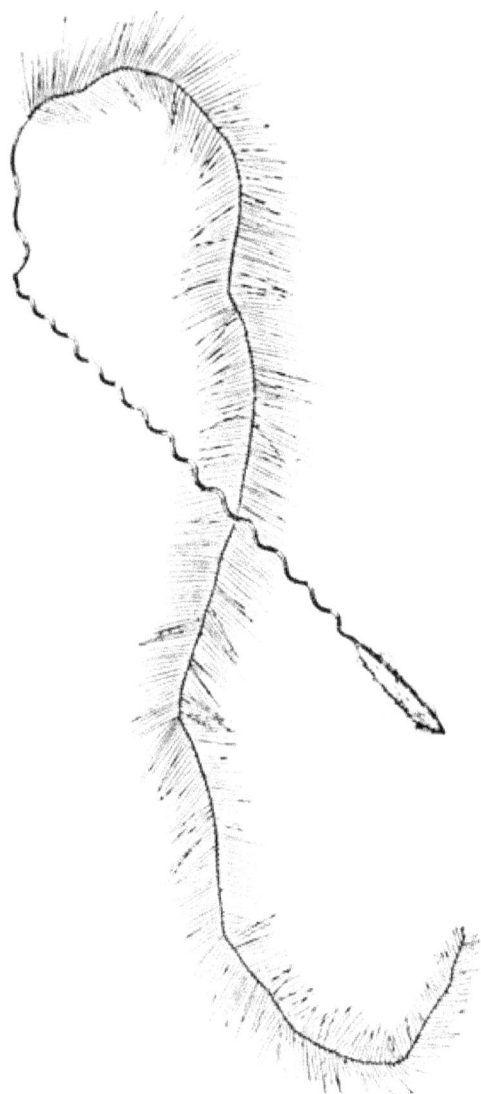

STIPA PINNATA (FEATHER GRASS).

seen to wave its long horns as if they were the

antennæ of an insect, and to turn over and over
until it has progressed some inches from the point
where it was first placed. In *Avena elator* (the tall
oat grass) and *Stipa pinnata* the awns are bent
sharply just as they emerge from the flowers, the
part below the bend being like a corkscrew and
highly sensitive to moisture, relaxing and con-
tracting according to the amount of moisture in
the air, with the result that the seed travels along
the ground. By the help of the long awn it can
pass over small obstacles, such as stones or clods,
the movement resembling that of a lever.

I must here guard my readers against those
movements that are caused by some insect larva.
The so-called jumping bean imported from Mexico
is now so well known that it may be taken as a
type of these curious movements due, not to the
seed itself, but to the efforts of an imprisoned
insect, the grub of a small moth which passes its
larval stage inside the hard-shelled seed of a kind
of euphorbia.

In conclusion, we may glance at a small group
of plants that develop sticky glands for the purposes of dispersion.

That charming Alpine plant *Linnæa borealis*
has a pair of bracts closely adherent to the fruit
and these bracts are covered with stalked glands
of a sticky nature, so that when an animal, bird,
or even a passing moth brushes against the little
fruits they stick to the intruder and are thus borne
away.   Now it may perhaps occur to the thoughtful reader that the Linnæa seed-vessel, being part
of a growing plant, would not readily break off
with a slight touch, but it is another instance of
that consummate skill and arrangement that is so
apparent to the close observer.   In the stalk of the
little fruit there is a special separating layer [1]
(analogous to that of the falling leaf which we
noted in a previous chapter), and at this point the
fruit readily separates if the slightest pressure is
brought to bear upon it.   This example is typical
of what takes place in such plants as *Salvia
glutinosa* and *Plumbago capensis* and *Rosea*.   As a
contrast to these various modes of dispersion I
may mention those seed-vessels which are actually
buried by the plants themselves, such as the
ground-nut, ivy-leaved toadflax, and others.   We

---

[1] Called botanically an "absciss layer."

must bear in mind that these plants usually have
aerial flowers in addition to those matured underground, and that these aerial flowers produce fruits
which are subject to dispersion. We may therefore
conclude that the underground seeds are to ensure
the continuance of the plant when the ordinary
methods have perhaps partially failed. My
readers may each autumn find an endless source
of wonder and interest in the thousands of differing fruits and seed-vessels which may be obtained
in any hedgerow and field ; and by careful observation they may yet learn many new facts and be
ever adding to their store of knowledge by gathering and comparing the fruits and their dispersion,
as shown in the types sketched in this chapter.

Objects to collect and examine :—Fruits and
seed-vessels of martynia, burdock, forget-me-not,
agrimony, enchanter's nightshade, bedstraw, samara
of sycamore, ash-keys, pinus seeds, birch seeds,
dandelion and goat's-beard seeds, pods of sweet-
pea, broom, laburnum, and pansy. Seed-vessels of
balsam, wild oat, feather-grass, *Linnæa borealis*,
salvia, plumbago, ground-nut, and ivy-leaved
toad-flax.

CHAPTER XI

GERMINATION

" O Source unseen of life and light,
  Thy secrecy of silent might
    If we in bondage know,
  Our hearts, like seeds beneath the ground,
  By silent force of life unbound,
    Move upward from below."

T. T. LYNCH

CHAPTER XI

HAVING considered the processes which lead up to the formation of seed, we may now investigate the life-history of a seed and its various forms.

Like fruits, seeds differ much in their outward shape. In size alone we find a great contrast between the dust-like seeds of the orchids and the huge seeds of the cocoa-nut-palm, while between those two extremes we may note every gradation of size. In other respects, also, the seed offers no less variety of form and covering than the fruit, such variations having relation to the particular mode of dispersion and germination. The outer skin or coat of a seed, called the *testa*, offers a very interesting field of study, and such seeds as the poppy and *silene* with beautiful net-

work, the *bignonia* and *pinus* with membraneous
wings, the cotton-plant seed with long hairs, and
the *collomia* with hairs that are resolved into
mucilage when wetted, are all worth special study.
When a small portion of *collomia* seed is moistened
and placed in a microscope one may see the rapid
change being effected ; that which had been a hard

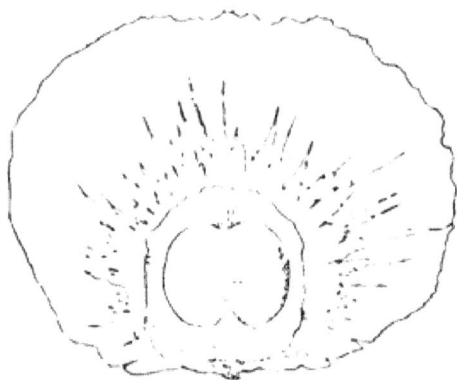

BIGNONIA SEED.

dry atom suddenly throws out coils of gum, like
watch springs, and a novice is led to ask, " Is the
thing alive ? " so full of motion does the object
appear.

We may regard a seed under various aspects.
As a special means of continuing the life of a
plant, one of its modes of reproduction, as a special
means of tiding a plant over a season that would

be fatal to its life in its ordinary condition of leafage, in the seed we have the germ of the future plant, a reproduction of its parent. This germ or embryo is lethargic or hibernating like many animals which exist throughout the winter in a dormant condition, yet still continue to be living vital bodies waiting for some special influence to come into play, and ready to resume all the activity of a growing organism. The construction of a seed is simple; inside the coat or *testa* we find the embryo with or without a special supply of albumen; if the seed is ex-albuminous, then we may expect to meet with thick, fleshy seed-leaves especially stored with this substance. The embryo contains all the essential parts of the plant, the root, stem, and leaves; the root in the seed state is called the radicle, and is that part of the embryo which usually points towards the micropyle; this radicle forms one end of the first shoot which comes out of a seed, the other end terminating in the stem or plumule. This first shoot is known by three names—axis, *tigellum,* or hypocotyle. The *tigellum* in many plants gives rise to a special structure; thus in the cyclamen it forms the tuber,

and the greater part of the "roots" of radishes
and turnips is due to it. In other instances it is
a mere collar forming a slightly thickened surface
between the base of the cotyledon and the radicle.
The *tigellum* is in reality a centre of growth, as
may easily be shown by cutting off an inch of the
upper part of a well-grown carrot and placing the
slice in a saucer of water; before long a crown of
young leaves will spring up and will continue to
grow and flourish as long as the plant food
contained in the slice is sufficient to maintain the
leafage. In botanical language we have thus been
growing carrot leaves from this *tigellum*.

The embryo varies very much in the relative
position of its parts. Thus the embryo of the
reed-mace is straight in the *tigellum* of the em-
bedding albumen. In contrast to this is the curved
embryo of the deadly nightshade and the spiral
embryo of the hop.

The seeds of the orange often contain two em-
bryos, which is rather a rare occurrence in the vegetable world. Before we can trace the future of these
parts we must attain a clear idea of the change
the seed undergoes when it germinates. In the

whole of our studies our attention has been drawn to no process so deeply interesting and yet so mysterious as that of the breaking into life of the seed. There are three conditions that promote the process of germination: warmth, moisture, and air. When these three conditions are present and the seed is healthy, growth begins, and its first stage is the absorption by the seed of moisture; this, combined with warmth and the oxygen of the air, sets up a change in the contents of the seed. We have already seen that seeds are of a dry and starchy nature, and in this condition they are in-

DOUBLE EMBRYO OF ORANGE.

soluble and unfit to be active plant food. The change that ensues results in this starchy matter being converted into sugar which is soluble; then the parts of the embryo begin to unfold, first the radicle and finally the plumule are developed.

In this early stage these parts live entirely upon the contents of the seed, just as a young chick is developed and nourished upon the albumen of the egg.

The temperature requisite for germination varies according to the species; those of us who possess gardens know to our cost at what a low temperature such plants as chickweed, bittercress, groundsel, and some of the speedwells grow; as long as the thermometer is above freezing-point these troublesome weeds will make their appearance in our flower borders. Sach's experiments on germination tend to show that wheat and barley begin to grow below five degrees centigrade, whilst French beans and maize germinate at nine degrees centigrade.

Some plants start into growth very quickly. Garden cress, vegetable marrows, and some grasses appear above ground a few days after they are sown, whilst other seeds, enclosed in a hard, woody seed-case, will require twelve months to germinate. This was the case with a seed taken out of a cedar cone brought from Mount Lebanon; I vainly watched for the young plant, and when a year had passed by the pot was thrown aside on a rubbish heap. Shortly after I was passing by and observed a fir-cotyledon growing on the heap, and this proved to be the long-desired young cedar-plant.

Seeds have the power to retain their vitality for years, especially those of the *Leguminosæ*, but I believe the stories of Egyptian mummy wheat germinating are scarcely to be believed. A good object-lesson upon this subject is furnished by a newly-made railway cutting; here we may always find growing upon the freshly-turned soil quite a crop of plants which have sprung from seeds that in the course of years have become embedded in the earth, it may be at so great a depth as to preclude the admission of air or prevent one of the necessary conditions of germination. When, however, the underlayer of soil is brought to the surface and exposed to light, air, and moisture, the seeds are able to grow.

To this we owe the richness of our railway-bank flora, and many a rare plant may be discovered there which cannot be found elsewhere in the neighbourhood. We will now in imagination conduct a few simple experiments that we may

17

learn something of the behaviour of seeds during
their early stages of growth. Each seed that we
thus study may be regarded by us as a type of
many others. First, then, we will sow, in a few
pots, about a dozen broad beans ; before doing so
we may notice on the seed the black stripe or

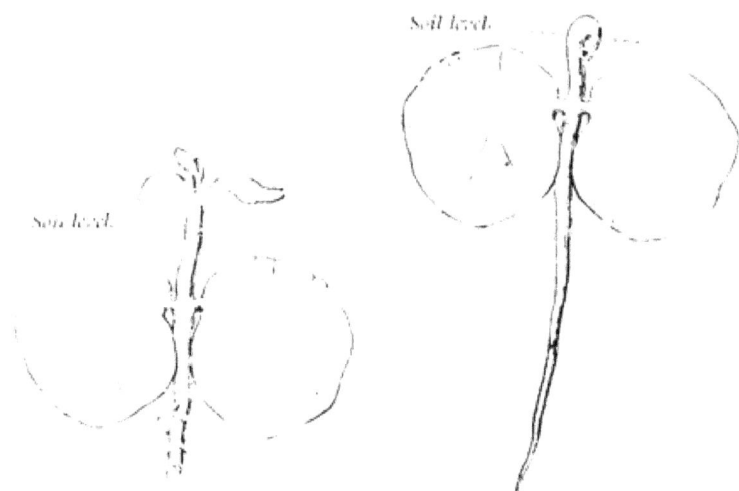

BROAD BEANS.

ridge known as the *hilum* ; this is the scar showing
where the seed was attached to the pod, and at
one end of it is the micropyle *(small gate)*. If we
remove the skin of the seed we shall observe the
two fleshy cotyledons or seed leaves, a tiny point
which is the rudimentary root, and, lying close to
the inner face of the cotyledon, the slightly curved

plumule. After the beans had been sown a few
days and carefully watered, we may take up two
or three for examination. At first we may only
see the radicle just emerging from the little hole
at the end of the *hilum*, but if we wait, say, eight
or nine days, we shall get a further development.

Before digging up our seed we will see if any
others are peeping through the soil. Yes, here is
one, just an arched kind of shoot, no leaves, only
the bow of the arch pushing up the particles of
the soil, so that the point of the shoot is clearly
still below the ground. Now, taking up a seed we
notice that the radicle has penetrated some way
down into the soil, and with a pocket lens we are
able to see a little higher than the tip of the root
quite a crop of delicate little root-hairs. The
cotyledons are still enclosed in the tough skin, but
the upward growth of the *tigellum* is acting on
them like a lever, and we can now plainly see that
it is this *tigellum* that, by its upward growth, is
penetrating the soil, and in so doing is drawing
the cotyledons from the seed coat. All this time
the delicate plumule is kept out of danger by the
arched shape of the *tigellum* and the folding of the

cotyledons. Leaving our seeds for a day or two
longer we find a further change. The plumule
has been carried up beyond the soil-level and has
begun to expand into leafage. It is interesting to
note how the curved *tigellum*, pushing through the
soil first, effectually guards the plumule from
injury arising from contact with rough particles
of earth; the cotyledons remain just below the
soil-level and we see that the *tigellum* is thickening and forming a distinct connecting branch
between the new shoots and the fleshy seed leaves;
these latter are full of plant food, and the plumule
is supplied from this storehouse of nutriment until
the first leaves are formed and are able to decompose carbon-dioxide for the nourishment of
the plantlet. The seed-leaves in this case do not
perform this function, but act simply as storehouses.

Our next seed example will be the familiar
mustard plant. These we may sow in two lots,
the first we only need to sprinkle upon some fine
soil and the second may be sown in a shallow drill
and covered with fine earth.

The first sowing will quickly germinate, and the

movement of the radicle which pushes out of the
micropyle may be understood by reference to the
appended diagram. In it we see the white thread-
like radicle emerging from the seed coat ; it turns
very quickly towards the ground and pushes
directly into the soil. Here I must direct my
readers' attention to one of those minute arrangements which, though apparently insignificant
enough if we fail to study the context, is really
an evidence of the infinite perfection, care, and

GROWING MUSTARD SEEDS.

wisdom of the Creator in even such a tiny detail
as the springing up of a mustard seed. As the
seed lies upon the ground, the lengthening radicle,
while it penetrates the ground, has a tendency to
force the seed into the air (as shown in the illustration), and were it allowed to do so the seedling
would soon shrivel up and die. This catastrophe
is, however, averted by the development upon the
radicle of quite a crop of fine white root-hairs :
these adhere closely to the minute particles of the

soil, and are thus enabled to counteract the force
exerted by the tip of the radicle; the latter pushes
through the ground without uplifting the seed.
This action can be watched and the growth of the
root-hairs observed by means of a pocket lens and
by the exercise of that virtue, most necessary for
all young naturalists –patience.

Returning to the seeds that were sown under
the soil, we find they have germinated ; the radicle
is pushing downwards, and just above the soil-
level we may see the short curved *tigellum*. This
very quickly straightens itself, and then we observe that the cotyledons have been drawn out
of the seed-coats and are displayed as two green
leaves, which in a few days will be an inch or two
above the ground, owing to the growth of the
tigellum. Here we get quite a departure from the
bean seed, whose cotyledons were *hypogean* (under
the earth), those of the mustard being *epigean*
(upon the earth). There is also another point of
difference; the mustard cotyledons are green, they
contain chlorophyll corpuscles, have stomates, and
so can perform all the functions of the normal
green leaf; thus they help at once to feed the

young plantlet by decomposing the carbon dioxide
of the air and forming starch, whilst in contrast to
this we learnt that the seed-leaves of the bean
were storehouses only. We are now sufficiently
acquainted with the functions of the seed to be
able to appreciate the variations of the *testa*, or
seed-coat. In numerous instances the spines,
prickles, hairs, and other growths on the surface
have, in addition to their use in dispersing the
seed, an essential purpose in holding the seed in
its rightful position. We will take cress as our
next example, since it may be regarded as a type
of all smooth seeds. Cress seed remains intact
until water comes in contact with it ; then it
becomes slimy by the liberation of a mucilaginous
cement from the outer coat layer ; this is, of
course, highly adhesive, and thus the seeds are
fixed firmly into the soil.

Another example is that of the little epiphyte
mentioned in our first chapter, *Tillandsia usnoides*,
or old man's beard. When the seeds leave the
capsule they are furnished with silky hairs, which
enable the tiny little structures to float through the
air ; they soon come in contact with the bark of

trees, and then the little hairs cling to the rough
surface.   In this position the seeds germinate, and

BEECH COTYLEDONS.

are held firmly in their place by the tightly-clasp-
ing silken strands.

Hardly any pursuit is more delightful than the
collecting and drying of seedling trees ; a ramble

through the woods in early summer will reveal
many specimens under or near the outskirts of the
foliage. Under the beeches we shall soon light
upon the nuts of last year coming up through the
moist, rotting soil, in the form of two broad, green
seed-leaves. As they often retain the dry, three-
cornered seed-husk upon them, we can easily see
that they are young beeches; otherwise, the
cotyledon leaves being so unlike the perfect form,
it might be rather difficult to distinguish the
species. These seedlings have germinated somewhat like the bean seed, the radicle has grown
downward, and the curved *tigellum*, pushing upwards, has drawn the cotyledons out of the seed-
coat. We may notice with surprise through how
small an aperture the cotyledons have been
pushed, and still they are uninjured, a fact that is
due to their being folded up like a fan in the seed-
husk. As soon as the *tigellum* reaches light and
air it straightens out, and the flat seed leaves, which
are at first of the palest green, soon deepen in
colour, and are working away preparing food for
the growth of the young plumule which springs up
from between the cotyledons, crowned with two

perfect young beech-leaves. This is all the baby-tree can do the first year. We can distinguish the second-year seedlings by their woody stem, brown leaf-scales, and silken-fringed young beech-leaves.

We shall not find cotyledons on the young oak, horse-chestnut, or sweet-chestnut seedlings, because these remain normally below the ground (hypogean), forming a storehouse of nutriment for the young tree. It is interesting to watch the growth of an acorn when placed in damp moss in a saucer. After a few weeks the acorn will have absorbed water, and the leathery seed-coat will burst at the pointed end; through this rent the radicle will protrude, fibres will be found growing upon the root, the *tigellum* is thick, and just where the stalks of the cotyledons are joined to it the plumule emerges as from a sheath. The plumule is in

ACORN.

no hurry to develop leaves: its first growth is provided for by the rich supply of food within the acorn. If, however, we look carefully at its little stem, we shall observe upon its surface a few scattered scales, each with a rudimentary bud in its axil. When the shoot has attained a height of three or four inches it develops its first green leaf,

HORSE-CHESTNUT.

and by the end of its first summer about six will have been formed. A collection of these seedling trees, carefully dried¹ and neatly arranged in a blank book, with the English and Latin names to each, a note of the age of the seedling, the

¹ They merely need to be placed between sheets of blotting paper, which should be dried daily and kept in a press or under a weight for a few days until the specimens are fit to be placed in a book.

spot where it was obtained, and the date, will in time form a pleasant memento of forest rambles, and, probably, may lead to further studies of a similar kind.

To make the collection complete there should be some seedlings of the other great division of plants, namely, the plants with one seed-leaf (monocotyledons). A few date-stones will supply these specimens; they should be sown in moist earth and placed either in a greenhouse or on a sunny window-ledge, where their growth can be watched.

Their germination is quite different from that of the other seeds we have described, and if a number of seeds are sown the different stages can be seen as in the accompanying figure.

One long cotyledon is pushed out from the seed, the free end is like a sheath. The part nearest the seed forms a structure resembling a rolled-up stalk: from the former roots are developed, whilst from the rolled-up stalk or sheath grows the next formed leaf, and each successive leaf is sheathed like its predecessor. This arrangement can be well seen in young growing grasses which can be

taken to pieces and examined. I shall conclude
this chapter with a brief reference
to the spores or so-called seeds of
ferns and mosses.

These are essentially different
from the seeds that have formed
our study in the earlier part of
this chapter, they do not contain
an embryo. Let us first notice
fern-spores, which we shall find in
abundance at the back of maiden-
hair and other fern fronds ; they
are contained in little brown
patches known as spore cases
(sporangium, from *spora*, a spore,
and *aggeion*, a vessel . If we collect
some of these and sow them on
some very fine damp earth, keeping it at the same time shaded
and warm, the spores will soon
germinate. We shall not find a
radicle this time as the result of
growth, but in its stead a flat
expansion of green tissue (pro-

YOUNG DATE-
PALM.

thallium, Gr. *protos*, first, *thallos*, a branch)
growing upon the earth like an exceedingly
delicate leaf. From the underside of this green
film a few very fine root-like hairs (rhizoids, Gr.
rhiza, a root) are developed; very soon with a
microscope we shall be able to discern upon the
surface of this structure a few little projections.
In one of these is developed a flask-shaped mass
of cells (archegonium, Gr. *archegonos*, first of a
race), in the other (antheridium, diminutive of Gr.
anthera, an anther) some minute bodies (anthero-
zoides, Gr. *anthera* and *zoid*, a minute life) with
tails; these escape from the covering and wriggle
about very much like tiny animalcules until finally
they come into contact with the flask-shaped opening before mentioned. These tailed structures are
something like pollen grains in their function, only
they differ from pollen grains, which are passive,
by being endowed with the power of motion; the
result of their fusion with the flask-like body is to
fertilise the germ cell (oospore, Gr. *oon*, an egg) in
that structure, and from the germ cell so fertilised
is developed an embryo from which at once springs
the young fern plant. The first leaf grows from

the upper part of the embryo and from the lower part is developed the "foot," a little connecting-link between the green prothallus and the baby fern which serves to nurse the little plant until two or more leaves have been produced ; the roots also grow from the same part of the embryo. I imagine that fern spores could be grown and watched through all their various stages even by those of my readers who dwell in towns, as a bell glass would maintain the requisite dampness and shelter the young ferns from smoky air.

Lastly I will describe an even simpler form of spore development. At any season of the year we may find the capsule fruit of mosses (Calyptra, Gr. *Kaluptra*, a veil), a very common one being the hair moss (Polytrichum, Gr. *Polutrichos*, having much hair), borne upon long wiry stalks. Inside the capsule we shall find a large quantity of small greenish bodies ; these are the spores, which of course fall out when the spore-case is blown by the wind, and being light are easily carried away and at length find a resting-place in some damp nook or shady bank. In such a place they find the conditions necessary for their germination, which is not unlike the same process in other seeds and spores we have studied. The result is very simple. A fine, silky, thread-like body (protonema, Gr. *protos*, first, and *nema*, a thread) is developed; when this has attained a fair size, a little moss plant begins to grow upon its surface exactly as we see a bud grow upon a tree-branch, and it is upon this moss plant that the organs of reproduction are produced. We have now come to the end of our study of seeds.

An endless source of interest to the student of nature is opened up to view by carefully observing the beginning of all vegetable life, and the seed or spore of the commonest weed or fern will teach us lessons that should ever make us mindful of the wonderful mystery of life and its genesis.

Objects to collect and examine :—Poppy, silene, and collomia seeds. Examine *tigellum* of cyclamen, radish, and carrot. Sow broad beans, mustard, and cress seed. Collect seedling trees. Sow date-stones. Examine fern and moss spores.

CHAPTER XII

THE PHYSIOLOGY OF PLANTS

18

"Lo! on each seed, within its tender rind,
   Life's golden threads in endless circles wind;
   Maze within maze the lucid webs are roll'd,
   And, as they burst, the living flames unfold."
                        ERASMUS DARWIN, *The Botanic Garden.*

IN this chapter I will endeavour to present to my readers a concise view of the nature and method of the various processes that go on continually in the growing plant.

These processes were incidentally referred to in our examination of the character of the various organs of the plant. Thus, in dealing with the root, we spoke of its physiology so far as concerned the absorption of water by its root-hairs. In the leaf, we touched upon the correlation between the shape and arrangement of the leaf tissues and the part the leaf plays in the economy of the plant. The physiology of the

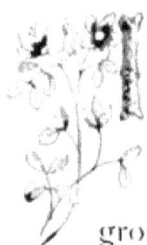

reproductive organs, again, we briefly explained in connection with their natural history.

In order to arrange our studies systematically, we may divide the physiology or function of plants into groups, and, taking each group separately study their effect on the plant.

We may then divide the functions of plants into

Nutrition,

Assimilation, and

Reproduction.

The first teaches us how a plant feeds and what it feeds upon; the second, how the food is prepared by the plant so as to enable it to use this food for growth and to store some of it away for future use. The third group deals with the various means adopted by plants for multiplying and increasing the species.

Plants, like animals, must *feed* and *breathe* in order to live; the food of plants, however, differs from that of animals in being more simple and elementary.

Plant food is of two kinds, water and gas. Water is an actual necessity to the plant, both as a direct food and as a medium to convey inorganic food. If we burn some wood to a white ash and then analyse it, six inorganic elements will always be found—potassium, magnesium, calcium, iron, phosphorus, and sulphur. These substances have been proved by experimental water-culture[1] to be indispensable to plant-life ; others are found in larger or smaller quantities, but they are not, judging by experimental tests, essential to plant life. These inorganic elements do not enter the plant as such, but in the form of salts dissolved in water ; the phosphorus and sulphur as phosphates and sulphates. Exactly how these salts and other elements are absorbed will be best learnt from a simple experiment.

[1] Testing the effect of plant food by water-culture is carried out in the following manner. Six large jars are filled with distilled water. In No. 1 all the six elements above mentioned are placed in small quantities, so as to form a weak solution. In No. 2 only five of them are added to the water, and in each succeeding jar one element is left out. A seedling plant which has been germinated on damp sand is suspended in each jar in such a manner that the leaves are in the air and the roots in the water without the seed touching the liquid. The growth of the young plants is carefully observed, and the result is found to be that No. 1 will grow and flourish, finding all its needful food in the water, whilst the rest of the seedlings will show plainly by their feeble and starved condition that, the food elements being absent, they cannot build up their stems and leaves, and must eventually perish.

We must first provide a large glass jar three parts full of clear water. Then a lamp chimney, to the bottom of which a piece of membrane (which any butcher will supply) has been affixed, should be partly filled with water coloured by sulphate of copper, and then suspended in the glass jar. Through a cork fitted to the top of the lamp chimney a long tube should be inserted. The fluid in the lamp-glass will be seen to rise in the tube shortly after the experiment is made, and the clean water in the large jar will become slightly coloured.

This experiment teaches us that liquids have the power of

TRANSFUSION DIAGRAM.

passing through a membrane; this power is known as diffusion, or *osmosis*. Further, we notice that the clear fluid passes into the coloured water more rapidly than the heavy coloured water passes out.

Now the fine *root hairs* of a growing plant are

membranes, having the same property as the membrane we placed on the lamp shade; inside the root hairs there exists heavy dense cell sap, outside are the films of hygroscopic water containing (dissolved) inorganic salts, and this water passes in through the membrane of the root, whilst a very little of the cell sap passes out into the soil, the quantity passing in being greatly in excess of that which escapes.

When once the crude water of the soil is inside, it is soon passed along to the stem and leaves by the pressure of more water coming in, and by what is called *capillary power*; this power we may easily see if we dip a fine tube into water, when at once the water will rise up some distance into the tube. I have pointed out that plant food is gaseous as well as aqueous.

Oxygen is absorbed by the root very freely from the soil, and, therefore, farmers and gardeners frequently plough and stir the soil of fields and gardens so that the roots may obtain a supply of this needful gas.

Let us now endeavour to see how the gaseous food is taken into the plant. In order to do so

we must remember that the gases necessary for
plant food form part of the air we breathe; this
air is made up of two-thirds nitrogen, one-third
oxygen, with a small and varying, but always
present, quantity of carbon-dioxide, and of these
the latter is the most essential to the life of
plants.

We have learnt in our study of the leaf how it,
by the aid of the green chlorophyll granules, and
under the influence of sunlight, absorbs this carbon
dioxide and effects certain changes in it. One of
the most essential elements in the growth of plants
is *nitrogen*; this we have just seen constitutes two-
thirds of the air we breathe, but the plant is unable
to make use of it in this free form; that is to say,
although the leaf can freely absorb carbon-dioxide
it cannot absorb nitrogen; it has to be taken in
by the roots of ordinary plants in the form of
nitrates, that is, in conjunction with some other
element. There is, however, an important exception to this rule; for what are called the
insectivorous plants have the power to absorb
nitrogen under certain conditions. These will be
explained in the succeeding chapter. We can

now summarise the processes of nutrition. The
roots absorb water containing earthy salts as well
as oxygen gas. The leaves absorb gaseous food
in the form of carbon-dioxide, and I may add
sometimes water vapour. There are two simple
experiments that my readers can make which
will prove these statements, and will give them
a greater interest in the somewhat dry details of
vegetable physiology. Our first experiment to
show the absorptive power of roots is taken from
Sir Joseph Hooker's Primer on Botany.

"Take up three plants of the buttercup carefully
by the roots; leave one (No. 1) on the table; place
another (No. 2) with its roots in water; hang the
third (No. 3) upside down over a tumbler of water
with a few of the leaves in the water, but the root
exposed. In due time No. 1 will have faded;
No. 2 will be quite fresh; No. 3 will have the
parts not in the water faded. No. 1 shows that
water contained in the plant has evaporated from
its surface; No. 2 that the water has been absorbed
by the root and conveyed to the leaves; No. 3
that the immersed leaves have not supplied the
other portions of the plant with water."

The second function, assimilation, depends upon several processes that together go to make up the work of digestion and preparing plant food. These processes are transpiration, respiration, and evolution of oxygen; the latter process is associated with the feeding of the leaf—that is, the absorption of carbon-dioxide. This compound gas is under the influence of sunlight, and by the agency of the green colouring granules, decomposed into carbon-monoxide and oxygen; the latter is eliminated, whilst the carbon and a part of the oxygen is retained, and with the absorbed water is converted into material that the plant can use for the purpose of increasing its structure.

By a very simple experiment we can prove the escape of oxygen from the foliage of plants. A few sprays of such leaves as laurustinus, bay, arbor vitæ, and maiden-hair fern should be tied firmly to a piece of stone. We should have ready a soup-plate, a glass shade, and a tub full of fresh spring water (one large enough to allow the shade to be held upright under the water). When all is ready, place the bunch of leaves and stone in the glass shade held horizontally, and gradually sink

it under the water till the shade is quite full;
place the soup-plate at the open end where the
shade is, and slowly raise the glass until it is
upright, and then it can be lifted out and placed
on a table in a window where the sun or bright
light can reach it. The bubbles of oxygen will
soon begin to form along all the edges of the
leaves and the jewelled effect of the bouquet will
be very curious and beautiful. It is hardly needful
to say the stone is simply required to keep the
group in an upright position. By the following
day there will be a large bubble of oxygen collected in the upper part of the shade, eliminated
from the leaves by the aid of chlorophyll and
sunlight.

These changes resulting in assimilation are
always in correlation with the process known as
transpiration. The root is continually taking in
fluids charged with inorganic salts; these are
by the water conveyed to the leaves by means
of the network of veins, which we know by the
term fibro-vascular bundles. These, as we may
see in skeleton leaves, traverse the entire substance
of the leaves where the salts are used up in the

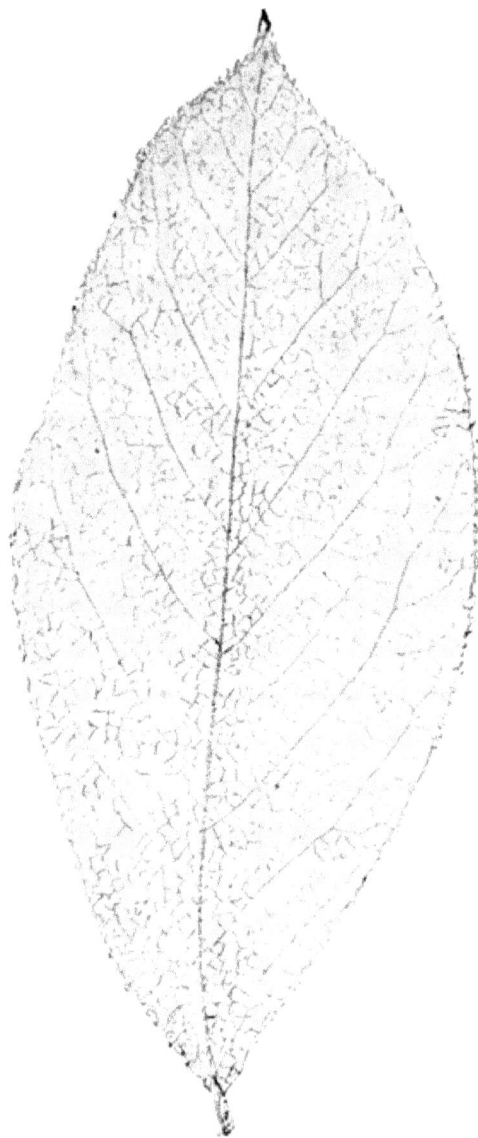

SKELETON LEAF.

constructive work of the plant. The water is not all wanted ; part of it passes off in the form of vapour. Transpiration, then, is the passing off of this water.

We can easily see this process going on if we place a few tropæolum leaves in a cool tumbler, and then expose the tumbler to sunlight. In a short time the sides of the glass will show a film of moisture due to the transpiration of the leaves. This process takes place more freely in a warm temperature than in cool conditions ; consequently, in hot weather there is rapid transpiration, and as the water is parted with more cell sap passes into the leaves and stems, and so the plant is kept cool. We can now see the great use of the little pores known as *Stomates ;* these are found mainly on the under surface, and it is principally through these pores that the leaf transpires.

We must now carefully note the fact that all growing parts of the plant take up oxygen and give off carbon-dioxide. This power which is common to all life is known as *respiration.* It is a process that cannot be observed in daylight in green plants because this respiration is feeble, and also because the opposite power of assimilation is so strong that the action of breathing is obscured. In the absence of sunlight, however, it can be observed, as also it may be traced in connection with parts of the plant other than the green leaves. Seeds, for example, during their earlier growth (germination) give off carbon-dioxide freely by respiration. This we can prove for ourselves by taking a large glass jar holding about two or three quarts; fill this about half full of beans that have been well soaked in water so as to swell them and induce them to commence germination. Close the jar with a tight fitting cork; after six or seven hours the presence of carbon-dioxide may be easily seen. Have ready a small phial of clear lime water, and with a piece of twine let this down into the jar without spilling its contents; allow it to remain there some minutes, keeping at the same time the top closed with a handkerchief. We shall see that the clear lime water will after a short time become cloudy or milky; this is due to the carbon-dioxide, liberated by the seeds, forming chalk with the calcium of the lime water, the chalk being

insoluble and easily seen. Now take out the
phial and let it stand, well covered, when the
chalk in the form of a fine precipitate will be
seen at the bottom of the phial. If desired, a
second experiment can be made with the same
jar by lowering into it a lighted taper; we shall
find it will go out owing to the presence of the
carbon-dioxide; as this gas does not support
combustion our lighted taper is quickly extin-
guished.

We can see from these experiments that respiration goes on in the growing plant and that this
process is independent of chlorophyll. It is an
essential part of the life of all plants, and my
readers who may perhaps wonder why it is
that two such opposite processes as I have
described are both carried on in the plant must
remember that in the main the feeding process
which depends on sunlight and the presence
of chlorophyll is carried on in the daytime,
whilst respiration is practically counteracted
in the daytime by the vigorous intake of carbon-
dioxide. At night when the rays of light cease
and no longer enable the plant to feed, the respira-

tion is evident. Briefly, we learn that *in light* the plant gains in weight, whilst *in darkness* (by respiration) it loses. The green plant can only construct growing material out of simple substances in light, having no power to do so in the dark.

Heat is just as needful to plant-life; it must be above freezing point, and a somewhat high temperature is necessary to set in motion all those chemical processes that I have briefly described.

At a low temperature the work of assimilation and other processes are arrested; on the other hand, a rise in temperature increases the activity of these processes.

We now come to the third function called reproduction. We have seen in connection with the food of plants how they convert inorganic material into organic. This one fact is significant of the great office of plant-life in nature; animal-life could not exist without its help. Plant-life may be said to prepare the food of animal-life, and retain that balance of gases in the atmosphere necessary to healthy respiration. How important then it is that all kinds of herbs, trees and plants

should multiply and be fruitful, life of any sort is
of limited duration, and subject to all the vicissi-
tudes of accident, constitution, and climate, and
so we find that plants have been endowed with
wonderful powers of reproduction in order that the
earth may be constantly clothed with vegetation,
necessary for the life of man and all animal
nature.

By reproduction I want my readers to clearly
understand the power possessed by the individual
plant to multiply its kind or species; and this
power is carried into effect in a variety of ways
in different species. These various methods of
reproduction then will occupy the concluding
pages of this chapter. The protoplasm (or life
principle) of any individual plant is endowed
with the power of giving rise to an entirely new
individual. This is accomplished in one of two
ways. In the first by cells forming a part of the
plant, but yet not specially modified for the purpose of reproduction. This mode of increase is
known as vegetative reproduction. We will illustrate it by two examples widely apart. Many
lowly plants like protococcus the bright green

substance which so beautifully colours tree trunks in moist situations) and yeast, are formed of one cell only, and when such cells attain their full size they simply divide into two or more cells which grow, and finally attain maturity when the process is repeated.

The other example is that known as the strawberry "runner," this, as we know, is only an elongated stem bearing at the end a bunch of leaves, and from the base of the leaves a few roots, the whole being a new plant which may be removed from the parent and grown in some other place.

These, then, are examples of vegetative reproduction, and my readers can discover for themselves many other instances in the garden.

The plan of propagation by "cuttings" is simply the gardener's practical application of vegetative reproduction.

The second mode of increase is by special reproductive cells, which are set free by the parent plants and become new individuals. The second mode is common to all plant-life, and in it two distinct processes can be observed. We often see on a decayed pear or apple a patch of brown mould (mucor). If we examine it with a lens we see a little forest of tiny erect stalks, and upon the apex of each is a round ball containing reproductive cells, each of these, which are called spores (the spore-case being called the sporangium), contains protoplasm, which is endowed with the power of giving rise to a new individual mould.

This process is typical of what is common to ferns, and many other cryptogamic plants, and is called *asexual reproduction.*

The second form is that in which two such spore-like organs as we have noticed in the mould, fuse together and form a spore capable of giving rise to a new plant.

This is known as *sexual reproduction,* and is dependent upon the fact that the protoplasm of either of the two organs is incapable of giving rise to a new individual plant, and that they must come in contact and fuse organically before a new plant can be formed. This process of fusion I have in an earlier chapter described as fertilisation. The pollen grain, the fertilising agent, is one of the reproductive cells, and the other, the ovule, is the cell that has to be fertilised. After this there is the subsequent development of the ovule into the seed, and in this seed we may recognise a plant in embryo endowed with powers not possessed by its parent, that enables it to resist extremes of heat and cold which would result in many cases in death to the parent plant. By way of experiment some seeds have been subjected to 40 degrees of cold, and yet have not lost their germinating power, whilst, on the other hand, it is known that seeds of some plants growing in sandy deserts lie baking in the sun for many months in a temperature of over 70 degrees, and yet begin to grow as soon as moisture reaches them.[1]

Things to be observed or collected :—Experiments to be made in order to show diffu-

[1] From "Nat. Hist. of Plants," p. 554 : "It has been proved experimentally that seeds which have been deprived by calcium chloride of as much water as possible are not killed even at the boiling point of water." Careful experiment has shown that there are three stages of activity in the life and work of a plant (1) A *minimum or zero*, at which the processes are just possible ; (2) a *medium stage* or *optimum point* where the activity is the greatest ; and (3) a *maximum stage of heat* where *growth is arrested*. So that we learn that plant-life can suffer from too high a temperature as well as that which is too low.

sion, transpiration, and respiration, collection of oxygen from water bouquet. Carbon-dioxide from germinating beans. Observe –

Blue mould on fruit.

Strawberry runner.

Rooted cuttings.

Stamens and pistil of any flowering plant.

CHAPTER XIII

INSECTIVOROUS PLANTS

.

" Beyond, the moorland has its wealth
    Of pink and purple, blue and gold ;
Heather and gorse, whose breath gives health,
    And ling. a hive of bees that hold ;
And when there's moisture in the brake,
    The clammy sundew's glistening glands
'Mid carmine foliage boldly make
    Slaves of invading insect bands."

INSECTIVOROUS PLANTS

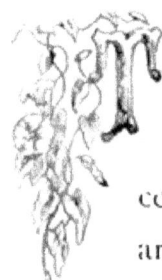

THE statement in our previous chapter that the leaf has no power to absorb nitrogen, has to be received with certain exceptions. These exceptions are discovered in a large group of plants, having little or no botanical relationship, and widely separated as regards their geographical distribution and habit of growth. The term insectivorous (insect-eating) has been applied to these by eminent botanists who have studied their habits and mode of growth. We may, as a preliminary to our study, summarise the main features of these interesting plants, because I wish my readers to see in them an extension and elaboration of the various processes we have tried

to investigate in plant-life, and not a mere description of a few vegetable wonders.   Rather would I
point out that in studying these deviations from
the ordinary type, as elsewhere, the young botanist
should try to arrive at some explanation of these
peculiarities, bearing always in mind that every
part of the plant is created for some special purpose.   This train of thought, if brought to bear
upon our botanical study will prevent our regard-
ing the contrivances of these insectivorous plants
as mere freaks of nature, which appears to me
to be a low and unworthy view to take of such
delicate and wonderful structures.

Occasionally, it is true, we meet with monstro-
sities, in the formation of which we fail to see any
hidden purpose ; but even here by careful observation we shall probably be able to perceive that it
is the result of some injury or the accompaniment
of disease from which plant-life is no more free
than animal-life is.

Let us now trace the features that are common
to the plants which form the subject of this
chapter.

Perhaps their most interesting function is that

of catching and retaining insects. This is accomplished in various ways, by viscid fluids which imprison small flies, as in the leaves of the sundew and other plants; by movements in the leaves, as in the Venus fly-trap; by a combination of both, as in the butterwort; or by special pitfalls and traps, as in the pitcher plants, sarracenias, bladderwort, and cephalotus. Having caught their prey, these plants dissolve it by means of an acid secretion; the dissolved animal-life is then absorbed and appropriated for the purposes of vegetable growth. Not all these processes are carried on by insect-eating plants. In some, for example, the secretion of dissolving acid is not very apparent, in others the absorbing glands are not fully developed; but, briefly, the above features are those possessed by this singular class of plants, and there is every reason to believe that powers of this kind are more widely spread than is usually supposed.

We will now notice a few types in detail.

The sundew (*Drosera rotundifolia*) is the pretty and poetic name of a plant which may often be found on boggy moors. It is barely an inch in

height, a mere rosette of leaves shaped like a
battledore, radiating from a very short root stock,
and bearing, in early summer, a central flower-
stalk from four to six inches high, furnished with
a few tiny white flowers. The whole plant lies

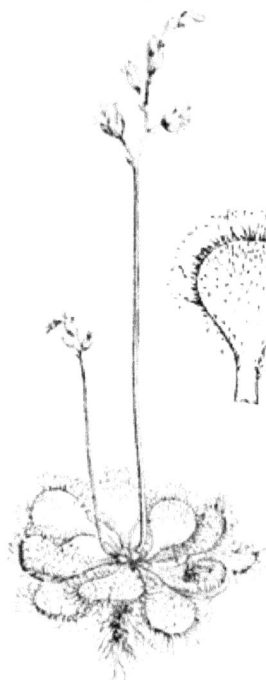

close to the ground, and is
often embedded in bog-
moss, and, were it not for
the bright colour of the
leaves [1] and their sparkling
dewy effect, it would be
a difficult plant to find.
With the naked eye we
can see that the leaves
are covered with hairs, and
a lens will show still more
plainly that these hairs
have each a club-like
end bearing a gummy

SUNDEW.

fluid, in appearance not unlike glycerine. These
globules of fluid sparkle in the sun; hence the
name of sundew and the botanical name of
drosera, from the Greek "*aroseros*," or dewy.

[1] On sunny heaths they are often of a rich crimson tint.

Leaves with glandular hairs are not rare amongst our wild plants, and if this was the only character that the sundew possessed it would not be specially noticeable. It is, however, the unusual structure and behaviour of these hairs that claims our notice. The term tentacle is a not inappropriate one to apply to these "hairs." A leaf of sundew, with all its tentacles standing out at different angles from the surface of the leaf, and each point armed with a drop of viscid fluid, is an effective arrangement for catching insects. The bright glistening drops are a fatal attraction to flies, gnats, and other small insects. When they alight upon the points of the tentacles they soon find that they are held prisoners. In their efforts to get free they entangle themselves more and more on the slimy points of the treacherous hairs. If we watch the tentacles after a fly has been caught, it will soon be seen that the hairs are bending over and closely pressing down the wretched captive. This folding over occupies four or five hours from the time the capture is made. The glands also begin to give out an increased amount of gummy secretion, and this

flow kills the insect by stopping up its breathing
pores, so that literally it dies of suffocation. The
fluid not only increases in quantity, but becomes
acid, and its effect is to dissolve the insect and
render it soluble ; the dissolved parts are then
absorbed by the glands and digested. This
interesting process can be watched quite easily
by carefully taking up a few plants of sundew
with some of the bog-soil and moss in which they
were growing and placing them in a glass dish,
where they will continue for months in perfect
health if kept very wet and covered with a bell
glass.

I once lighted on some magnificent sundew
growing on boggy land near Woolmer Forest.
Whilst taking up some roots of it I was persistently attacked by a stinging fly, and, my hands
being occupied, I could not well defend myself.
Happily the sundew acted a friendly part ! I was
carrying a tuft of it in my hand when, looking
down, I saw my tormenting fly was securely caught
upon its leaves. Somehow one always feels com-
passion for the unfortunate, and I confess I tried
to rescue the captive, but the creature's wings and

legs were already so glued together by the viscid dew that it was impossible to release it, and I realised more than ever how effective the sundew is as a fly-trap.

In transplanting specimens of drosera great care should be taken that the leaves are untouched, else, being sticky, they will cling together and lose their delicate beauty. Every few days the plants may be fed, and happily they are quite willing to accept very minute pieces of raw beef, so that flies need not be sacrificed in the cause of science. The little "beafeater" must not be fed a second time until the hairs have uncurled and the leaf has fully expanded, showing that the last meal has been digested. I have kept a large pan of sundew in great beauty for about four months in summer, and when the glass was taken off and bright sunshine lit up the jewelled leaves the effect was lovely, and a magnifying glass showed the structure of the leaves and the prismatic colouring of the dew-tipped hairs.

The Venus fly-trap is an exotic member of the insectivorous family. Its leaves are remarkably like an ordinary spring rat-trap A glance at the

drawing will show its formation. On the two lobes
of the leaf are a row of stiff bristles occupying the
precise position of the teeth of a rat-trap. The
inner surface of the leaves is of a reddish colour,

VENUS FLY-TRAP.

due to its being thickly covered with minute red
glands ; on each lobe there are three stiff hairs. If
a fly alighted on the leaf and walked across its
surface, it would touch one of these hairs, and no
matter how light the touch might be, the hairs are

so sensitive they would convey the signal to the hinge of the lobes, and they would instantly rise up and clasp the fly, eventually crushing it to death. Then would follow, as in the case of the sundew, the emission of acrid secretion and the absorption and digestion of the insect.

Insect-destroying plants are numerous in the vegetable world. They may be roughly divided into three groups, although there is no strict line of demarcation between them. First, those like the red lychnis and others, which, by means of sticky hairs, catch and kill small insects, an operation that, so far as we know, results in no special good to the plant. Then there are those, like the sundew, which catch, kill, and digest the insect for food; whilst the third group consists of plants which catch and kill insects, but have no digestive process. Decomposition of the captured insects takes place, but the absorption which goes on is simply that of the liquid products of decomposition, the latter process resulting from the insects being immersed in fluid. To this latter group belong the pitcher plants (*Nepenthes*) and sarracenias. These last are North American

plants of peculiar structure and appearance. The
leaf is folded and modified into a tunnel-shaped
tube differing in form in the various species.   In
all there is a kind of cap or lid to the tube, so
that rain is kept out.   In one or two species the

SARRACENIA FLAVA.

lid is so arranged that the mouth is exposed.   In
the bottom of these tubes there is usually a
quantity of somewhat slimy fluid. The inner
face of the lid and surface just inside the rim of
the tube is smooth, usually of a bright shining

colour and covered with minute honey-secreting glands, a most attractive lure for insects. Below this honeyed surface the character of the sides of the tube changes completely ; for, down to the fluid, it is covered with stiff hairs all pointing downwards. Now we see how the trap is set. The honey just inside the tube is attractive, and the insect feeding finds it very easy to descend the tube ; the smooth surface offers no foothold, and the downward pointed hairs prevent it from returning, until at last the insect becomes engulfed in the pool of water at the bottom of the tube. In this fluid insects generally accumulate, decompose, and become liquid manure.

In Georgia and North Florida these sarracenias are found in the swamps in large quantities attaining one to two feet in height, their great tubes half-filled with insects showing their value in tending to reduce the swarms of flies which abound in such localities. We can see from these characteristics of the sarracenia a link between the insect-eating plants which have a true digestive process and ordinary plants that obtain their food in part direct from the soil. The sarracenia

is simply making an attempt to collect nitrogenous food by the aid of its form and sweet
secretions; thus it lures on flies and other insects
to their doom, which to the plant means an

BLADDERWORT

increased supply of liquid manure for its nourishment.

Between the two types of insectivorous plants
and ordinary plants there are endless varieties.
The largest known species of " fly-catcher " is the

Roridula dentata of South Africa, which attains a height of six feet, with leaves similar to the sundew in character. So efficient are these leaves in catching flies that the Boers hang up branches in their rooms as fly-traps.

The smallest insect-eating plant is probably the bladderwort (*Utricularia vulgaris*), a rootless water plant with minute bladders on small thread-like leaves. The bladders only open inwards, so that when an insect pushes against the opening or valve it easily enters, and cannot get out again. The bladder contains water, but the insect quickly consumes the oxygen in it, and consequently dies, and when decayed its substance is absorbed by glands on the inner surface of the bladder.

PITCHER OF NE-PENTHES RAF-FLESIANA.

Perhaps the most attractive of the group of plants we are considering is the pitcher plant or Nepenthes. It grows commonly in Borneo and Ceylon. The pitcher is a direct development of the midrib of the leaf. It varies in size from the little thimble-like pitcher of *Nepenthes gracilis* to the large jug-like receptacles of *Nepenthes Rafflesiana* [1] and others, each capable of holding nearly a pint of fluid. The pitchers are furnished with a lid overhanging the mouth of the receptacle, this is kept open by a thick rim. This rim and the under-surface of the lid both secrete a sweet fluid which is attractive to insects, and from the rim and opening of the mouth a smooth surface directs the ill-fated flies to the sweet sticky fluid always found at the bottom of the pitcher, out of which they rarely come alive.

Another of our native plants exhibiting these insectivorous habits is the butterwort (*Pinguicula*). Like the sundew it is a mere rosette of radical leaves, having upturned margins and a very succulent pellucid appearance. These leaves are covered with glands which exude a viscid kind of fluid like that on the tentacles of the sundew. This natural birdlime catches and holds small flies, midges, and other tiny flying creatures, as well as crawling insects. The presence of these insects

[1] See Frontispiece.

on the leaf appears to stimulate it to further
secretion which must, of course, lessen the chances
of the insect's escape, and as a further barrier to
prevent its creeping away, the edges of the leaf
begin slowly to curve in-
wards, so that the caught
insect is imprisoned in the
folds of the leaf. The acid
secretion which now exudes
from the glands soon dis-
solves all the nitrogenous
and soft parts of the insect,
which are taken up by the
absorptive glands of the
leaf. There are many
other plants, of which I
have not space to make
mention, although they are
full of interest, as owing to

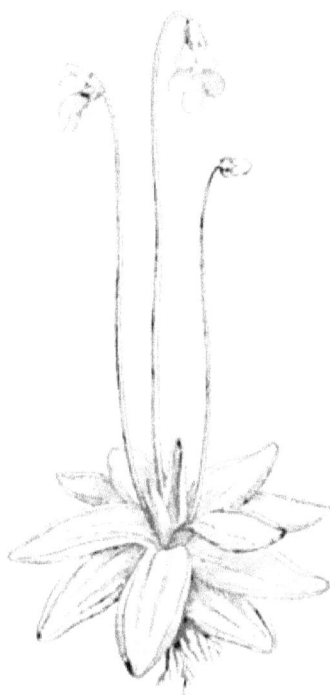

BUTTERWORT.

their curious structure, it is probable that insectivorous habits might also be ascribed to them.
The field of study is a wide one, and throws
much light upon the physiology of plants as
well as the relationship between the plant and

animal world. I would suggest to my young readers, as a practical means of knowing more of this subject, to try and grow for themselves the sundew, pinguicula, and sarracenia.

The two first can be found, as I have already said, on boggy moors in England, and the latter plant can be obtained from any florist. All can be successfully grown in a greenhouse or garden frame, and studying their growth and habits in this way will teach the young botanist far more agreeably than learning only from books.

At Kew there is always a fine collection of these insectivorous plants to be seen in vigorous growth, whilst at the South Kensington Natural History Museum (Botanical Department) there are some highly interesting cases illustrating the life history of these remarkable plants.

CHAPTER XIV

HABIT OF GROWTH IN PLANTS

"Some clothe the soil that feeds them, far diffused
  And lowly creeping, modest and yet fair,
  Like virtue, thriving most where little seen :
  Some, more aspiring, catch the neighbour shrub
  With clasping tendrils, and invest his branch,
  Else unadorn'd, with many a gay festoon
  And fragrant chaplet, recompensing well
  The strength they borrow with the grace they lend."

COWPER.

314

CHAPTER XIV

HABIT OF GROWTH IN PLANTS

MY readers have possibly noticed that in the previous chapters my aim has been to describe the various organs of a plant, and that I have tried to show not merely the botanical meaning of the many differences in the organs of allied species, but to point out also how these structures are adapted to help the plant to multiply itself. The object of this final chapter is to take a more general view of plant-life, and to give some idea of the different habits of plants; how in their struggle to grow and reproduce themselves they form such habits as tend to assist them in this effort, and also how entirely, in some cases, they differ from our ordinary conception of plant-growth.

We have already seen how beautifully plants are
adapted to the life they have to lead, how they are
specially fitted to grow in some particular place
and climate, and now I will ask my readers to
study with me certain of the varying habits of
plant-life.   A typical plant of an ordinary kind
grows, of course, in the earth, produces root,
stem, and leaves, and finally flowers, which are
the origin of fruits and seed ; by the latter the
plant is again produced, and by this circular
action the continuity of that particular plant is
maintained.

Let us now, in imagination, peep into a tropical
forest.   On its outskirts we shall see the prototypes
of our typical plant ; but inside there are also others
of quite a different aspect, and the first to attract
our attention would probably be the curious orchids
perched upon the tree-branches.   Their mode of
growth differs greatly from that of a normal plant,
for they are merely attached to the branches by
means of clasping rootlets, which do not in any
way extract sap from the tree to which they are
clinging.

The moisture they need is collected by the leaves

and hanging rootlets from the humid atmosphere
of the forest.   These plants that have acquired a
perching habit sometimes grow to an immense size,
and where they do so vegetable débris accumulates

PERCHING ORCHID.

about their lower leaves and roots to such an
extent that it serves to supply them with needful
food.

   This habit of growth is not confined to the lovely

orchids; mosses, lichens, ferns, and many other plants have acquired a similar mode of growth, and the various ways by which they attach themselves to the bearer plants would form an interesting subject of investigation. It is a not uncommon error to regard these perching plants as parasites, but this term is properly used for plants which actually feed upon the branches of the trees where they grow, and of course seriously injure the trees by so doing. The orchids, on the other hand, do not in any way injure the branch upon which they rest. Robert Louis Stevenson in one of his later poems has, with a poet's license, which in this case is contrary to fact, described the perching orchid thus—

> " For in the groins of branches, lo !
> The cancers of the orchid grow."

This inaccurate observation is, however, more than atoned for by the wonderful impression Stevenson has given us of the character of woodland strife, the ceaseless struggle for light and air which goes on in tropical forests.

In studying the parasites as a group of plants

associated by the same habit of growth, we are led to the conclusion that there is some difference after all in the morality of plants! Here, for example, we are confronted with a group of plants that differ entirely from those we have hitherto examined. The mistletoe, which is the commonest type, is certainly lower in the social plant-scale than the perching orchid, the latter with its leaves and rootlets being enabled to earn its own living, while the mistletoe sends its roots down into the soft sap of the branch upon which it is growing and there is no other name for it—steals its means of living and growing from the substance of the poor tree upon which it preys. It is true it does, in a half-hearted kind of way, assimilate a little gaseous food for itself, but the sickly metallic hue of its leaves is evidence that even in this respect it is shirking its proper duties of nutrition.

If we desire to study the curious habits of parasitic plants, the two examples referred to in a previous chapter, the clover-dodder and the yellow rattle, will afford good examples, the latter plant being easily obtainable in fields where the pasture is poor and scanty. Very

RAFFLESIA ARNOLDII.

curious are the modifications and contrivances developed by plants which have acquired this habit of parasitism, especially amongst such weird tropical species as *Rafflesia*, a huge parasite growing on the *Cissus* in Sumatra. When the leaves and flowers of the cissus have withered, then here and there a huge knob protrudes from the stem or root, and this grows in time to an immense stemless flower, measuring more than three feet across, its cup frequently containing as much as twelve pints of liquid, and the weight of the whole flower being said to be about fifteen pounds.

Differing a little in habit from the parasites are the *saprophyte* plants, which live on decaying vegetation. The little brown leafless orchid called the bird's-nest orchis is of this character, as well as the equally curious coral-root orchis. These plants, as well as many other parasites, are destitute of chlorophyll, and are therefore dependent on organic material for food ; this they obtain either as we have seen from living plants or from decaying organic matter. In their efforts to obtain a needful supply of light and air, some plants assume climbing habits, using as supports other trees and plants, to the very obvious disadvantage of the latter. We can well understand how, in a tropical forest, the weak-climbing plants strive to pass out of the shaded recesses and force their way to the tops of the slower growing trees, in order to obtain the share of light, moisture, and air which are essential to their existence. Very vividly has the late Mr. Louis Stevenson described such a scene in a tropical forest—

> "The hooked liana in his gin
> Noosed his reluctant neighbours in ;
> There the green murderer throve and spread,
> Upon his smothering victims fed,
> And wantoned on his climbing coil.
> Contending roots fought for the soil
> Like frighted demons ; with despair
> Competing branches pushed for air."
>
> *   *   *   *   *
>
> "So hushed the woodland warfare goes
> Unceasing ; and the silent foes
> Grapple and smother, strain and clasp
> Without a cry, without a gasp."

I may explain that the "murderer" alluded to is a species of fig-tree which, in its early youth climbs up the trunks of other trees, and by means of its clasping roots so constricts their stems that they ultimately perish.

In pleasing contrast to this phase of vegetable growth is the habit which indicates to us something of mutual help and co-operation. In the *Compositæ* we find many instances of a habit of growth that bears distinctly upon this "help-one-another" mode of life. A common daisy will serve as a type-flower of this kind. The little head is a colony of flowers, but so close is the association of its individual florets that it is usual to regard it as one flower rather than a distinct inflorescence composed of numerous separate and distinct flowers.

In order to understand the mutualism displayed by this little flower, we must remember that it is an insect-fertilised blossom, and, therefore, insects must be attracted to it. If we carefully dissect a flower-head we shall find first a ring of strap-shaped flowers on the outside, constituting the ray florets —these are imperfect ;[1] but placed side by side on the outer edge they become conspicuous ; then we find in the centre of the flower-head a number of tiny yellow flowers, each one containing stamens and pistils. What wee things they are, and if they were developed singly how inconspicuous they

[1] Barren.

would be! When, however, they are grouped side by side in the centre, and further, when the outer florets are of a different colour and shape, what a beautiful and symmetrical whole they make! Truly this is another rendering of the maxim, Union is strength. From a different point of view the arrangement is equally interesting. The white and pink tipped florets of the ray are not capable of bearing seed, and yet we see how they help those florets that are perfect by their attractive appearance; then at night or on a cold rainy day these same ray florets bend over and completely cover up the florets in the centre which are busy producing seed. My readers will find a rich field of investigation open before them in studying the flowers of the daisy family, and finding out for themselves how the florets are grouped together, and to what extent this principle of co-operation can be traced.[1]

Students will find the corn blue-bottle especially

[1] A single flower of the Heracleum giganteum would not be specially noticeable, but when hundreds of them are grouped together in a huge umbelliferous head they form a most striking object, as may be seen in the plate. I have often watched the swarms of flies, beetles, and bees visiting these attractive blossoms on sunny days, and the great umbels of seed in autumn showed how effectually the insects had carried out their work of fertilisation.

GIANT COW-PARSNIP (Heracleum Giganteum).

interesting ; the large outer florets contain no organs of reproduction, but still they are brightly coloured and highly attractive to bees ; the inner florets with their protruding stigmas and anthers, are much smaller ; they are the seed-bearers, and

CORN BLUE-BOTTLE.

cannot fail to receive pollination when the bee alights on the flower-head, allured by the showy outer florets, which apparently exist solely that they may draw insects to visit the unattractive flowers of the disc.

The direct influence of the separate parts of a

plant upon one another, and the very distinct habit of associating together that they **may** attain some end such **as** the **visits of insects,** leads us to consider **two other aspects of plant-life,** both of which **are so** full of interest that **no** botanical work **can** now be considered complete without some reference to the matter If **we carefully dig up a clover plant or a broad bean and examine the** little **root-** lets we shall **observe some small knobs or swellings upon them. These swellings are** only found **here and there on some of the roots,** so that **their** presence **is not a normal condition. Placing one of** these knobs **under a powerful microscope, we shall find** it **to be not ordinary root tissue but a** substance teeming with countless **numbers of rod-like or** rounded atoms which **botanists who have** investigated the **subject tell us are bacteria,** *i.e.,* **incon-** ceivably **small one-celled plants which are often** the cause of terrible **diseases. But some** of these **mysterious** organisms, **on the other** hand, **are** capable of beneficial results. **It has of late** been clearly **proved that leguminous plants having** these colonies **of bacteria on their roots** possess **the** power of assimilating the **free nitrogen that forms** such a large proportion of atmospheric air. When therefore a farmer sows his wheat in a field previously occupied by clover he finds the clover roots left in the soil contribute the best possible supply of nitrogen to the wheat crop. This seems a remarkable fact, since vegetable physiologists have hitherto insisted upon the fact that plant-life is unable to make use of the free nitrogen of the air. The other instance of strange habit is that of a symbiosis,[1] which exists between certain trees on the one hand and the threads of spawn of some fungi on the other. If the roots of the white poplar are examined minutely, quite a mantle of whitish threads will be found covering the growing point. It is said by that eminent botanist, Professor Kerner, and by others that, as the roots are developed from the young seedling-tree, they are enclosed in the meshes of the fungus, and that this particular fungus is always a close associate of the roots as they grow in all directions. This fact we can see when we dig up the roots, but the most striking part of the story is this, that between this

[1] A word meaning two plants living together and deriving mutual benefit.

fungus root and the roots of the tree there is an organic connection, a division of labour which results in the tree receiving from the thread-like filaments of the fungus (*hyphæ*) both moisture and certain food stuffs from the ground, whilst the fungus gets in return such organic food as the tree has produced by means of its green leaves. Such cases as these present to us a manner of growth that is akin to social habit, and, strange as the union may appear, the circumstance is by no means uncommon in the vegetable kingdom. Stranger still perhaps is the union that is sometimes to be found between plants and some member of the animal world, of which union I shall give an example. On one of the larger species of sea-anemones (*Anthea cereus*) are small yellowish spots, which at one time were supposed to form part of the animal itself. But now the spots turn out to be vegetable cells, which can be isolated and induced to continue growing after the death of the anemone. The yellow spots are small algæ, and are furnished with chlorophyll. We must not regard the algæ as parasites on the sea-anemone, because they split up the carbon-dioxide under the influence of sunlight, and by so doing supply the anemone with oxygen for respiration, whilst the starch formed in the protoplasm of the algæ passes by diffusion into the anatomy of the animal. The transaction does not end here; the algæ in all probability receives nitrogenous substances in return, so that there is a mutual interchange.

These are but one or two of the many wonderful phases of vegetable life, and I hope by thus briefly sketching a few of them my readers will be stimulated into a greater desire to explore God's marvellous works in nature. There is an endless succession of such wonders to be investigated, but in order to find them we need a careful spirit of observation, passing nothing by without trying to learn something of its life history. Every hedgerow is full of delightful problems which will reward the interested student. A single field has been found to contain as many as fifty different species of plants, and every month of the year will present a new aspect of life. In the early spring we have the germinating seed and the tiny growing moss. A little later the opening buds with their wealth of interesting points to study, then the unfolding of

the leaves and the gradual development of the
flower. Here and there a climbing plant will
engage our attention, its mode of climbing, its
modification of part or parts to enable it success-

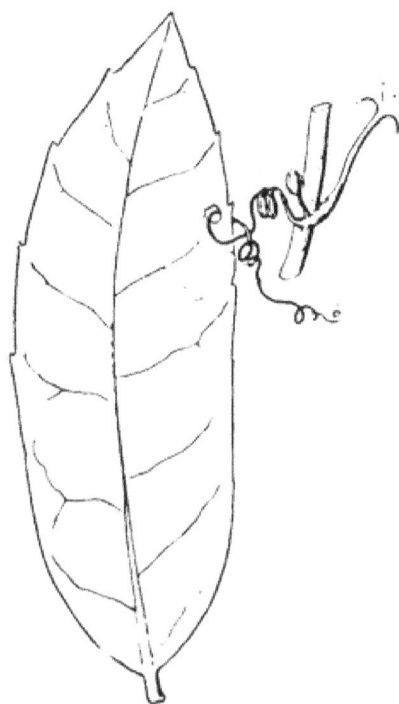

BRYONY TENDRIL.

fully to overcome difficulties, its acceptance of help
by the way—as in the case of a bryony tendril I
once came across which cleverly attached itself to
a minute hole in a laurel leaf—these and many

other items will interest us in our walks if we keep
our eyes open.

Then, as summer slowly passes away and autumn
approaches, the fruits will engage our attention ;
their forms and shapes and modes of dispersion
will afford ample subjects for study.

Winter, too, still brings its store of pleasure for
the young botanist. Nature is not dead—she only

TRICHIA THROWING OUT SPORES.

sleeps. Nay, unless there is hard frost and deep
snow the field for observation is just as wide and
the harvest as plentiful as at any other season.
Look on the old apple-trees and see what a host
of tiny plantlets there is there to glean. Here are
pale-green bearded moss and lichens, there a branch,
perhaps, lies on the ground dead and decaying,
under whose mouldering bark, if we have keen

eyes, we may discover tiny tufts of the *Myxetozoa*, whose capsules, under the microscope (and in some cases even with the naked eye) are seen to give off clouds of spores, actually thrown out by the active movements of fine waving threads, a sight never to be forgotten when it has been watched under favourable circumstances. Winter is also rich in its harvest of mushroom-like fungi; these will well repay a little study. We shall be led to note their form, colour, mode, and habit of growth, how they affect certain trees and soils, and the important difference of some kinds being eatable and others virulently poisonous; the mere book student can know very little of the keen pleasure enjoyed by those who thus think about what they see, and are ever adding to their stock of knowledge by personal observation. I may close with some true and beautiful thoughts by one [1] who is herself a reverent student of the book of nature.

"No pleasure is more sure and none less costly than that of watching day by day the signs of the coming spring; than the delight of seeing unex-

---

[1] Miss Blanche Atkinson, member of the Barmouth Branch of the Selborne Society.

pectedly the first primrose, and of finding that the
anemones and hyacinths are pushing their way to
the sunshine. Year by year the miracle of spring-
time, when the green leaves are shaken forth from
the hard bud is more miraculous. Summer after
summer the lilies are fairer, the wild roses more
exquisite, and on through the seasons the varying
pleasures succeed one another. These things never
pall ; and if the time should come when we can no
longer go out to the hills and woods to welcome
the spring and revel in the bounty of summer we
know that the past is not lost. The fair remem-
brance of the flowers of the field is safe in our
hearts, and will ' flash upon that inward eye which
is the bliss of solitude.'"

GLOSSARY OF SCIENTIFIC WORDS
USED IN THIS VOLUME

GLOSSARY.

A CLEAR definition of scientific terms involves an exact knowledge of several languages, and when translated into technical phraseology these definition often appear to me to be as difficult to a simple comprehension as the original words they purport to explain.

I have endeavoured therefore, in this glossary, to put scientific terms into plain words as clearly as was consistent with the facts, and not by any means to attempt a really exhaustive scientific definition.

A

Abscess—A term applied to a layer of separating cells.

Absorption—Taking in food by diffusion.

Accessory—Anything additional.

Acetic—Applied to an acid, sour.

Achene—A small dry indehiscent fruit with a leathery coat.

Adaptation—As applied to plant-life meaning the structure of the plant becoming most fitted to its environment.

Adventitious—Not developed in regular order.

Aerial—Inhabiting or existing in the air.

Estivation—The arrangement of the parts of the flower in the bud.

Albumen—Reserve material contained in the seed, analogous to the white of an egg.

Alchemilla—A genus of rosaceous plants with small green flowers.

Allium—The onion genus.

Altitude—Height.

Ampelopsis—A genus of climbing plants allied to the vine whose leaves are brilliantly coloured in autumn.

Anemophilous—Pollinated by the wind.

Animalcule—Microscopic insect life.

Annual—A plant whose duration of life is one season: *Ex.* mignonette.

Anthea—A genus of sea-anemones.

Anther—The dilated end of the stamen in which the pollen grains are developed.

Antheridium—The case containing the antherozoids in cryptogamic plants.

Antherozoids—The male cell, or active member in fertilisation of cryptogams.

Antirrhinum—The snap-dragon genus.

Antiseptic—Counteracting decay or putrefaction.

Apocarpous—Applied to the pistil when the carpels are distinct or when the pistil consists of one carpel.

Appendages—Something hanging or appended, extra.

Aquatic—Relating to water.

Araucaria—The generic name of the monkey-puzzle tree.

Archegonium—The flask-shaped organ containing the female cell in the cryptogams.

Arid—Dry and waterless.

Arillus—An out-growth from the funicle (or seed-coat).

Aristolochia—A genus of climbing plants with curious "prison" flowers which attract and retain insects.

Arum—A genus of poisonous plants with an inflorescence consisting of spadix and **spathe.**

Asexual—Not sexual.

Asparagus—**A** genus **of edible vegetables and** climbing plants.

Assimilation—**The** conversion **of crude food into** protoplasm.

Aster—The generic name of the **Michaelmas daisies.**

Avena—The generic **name of the oat.**

Awn—The beard of **barley and other corn.**

Axillary—Growing **in the axil of the leaf.**

B

Bacteria—Minute one-celled living atoms, the cause of most contagious diseases.

Bamboo—A giant grass.

Banana—The fruit of the genus Musa.

Bark—The rough external part of a stem.

Barm—Same as yeast.

Bast—The fibrous tissue between the bark and the wood of a dicotyledonous stem.

Begonia—A genus of plants with bright flowers and oblique or one-sided leaves.

Betula—Generic name of the birch-tree.

Biennial—A plant whose duration of life is two seasons : *Ex.* Beetroot.

Bifacial—With upper and lower sides structurally different *Ex.* laurel leaf.

Bignonia—A genus of flowering climbing plants.

Blade—The broad part of the leaf.

Bougainvillia—A genus of climbing tropical plants with bright pink bracts and small yellowish flowers.

Bulb—A dormant bud surrounded with fleshy scales.

Bulbils—Small bulbs.

Bunium—A genus of tuberous Umbelliferae, earthnut.

Buoyant—Light, able to float in air or water.

Button-wood—A term applied in America to the plane tree.

C

Cacti—A family of succulent plants usually devoid of leaves.

Caducous—Quickly dropping off.

Calceolaria—A genus of herbaceous garden plants with pouched flowers.

Calcium—An element present in all calcareous rocks.

Calyptra—The hood of a moss-fruit.

Calyx—The outer whorl of the flower or floral envelope, cup-shaped.

Cambium-layer—A layer of active growing tissue.

Campanula—A genus of Alpine and herbaceous plants with bell-shaped flowers.

Camphor—A drug obtained by dry distillation of the leaves and stems of Camphora officinarum.

Capillary—Fine and minute, hair-like.

Carbon-dioxide—Symbol CO_2. A gas existing in small quantities in the air, otherwise called carbonic-acid gas.

Carbon-monoxide—A poisonous gas whose molecule is composed of one atom of carbon and one atom of oxygen.

Carex—A genus of sedge-like plants.

Carpel—A pistillate leaf, one of the component parts of the pistil.

Caterpillar—The form of an insect after it is hatched, first stage.

Catkin—A spike of staminate or pistillate flowers usually pendulous.

Checkered—Outlined into a square-like pattern.

Chevaux de frise—An obstacle consisting of iron spikes set in a framework of iron.

Chlorophyll—The green colouring matter of leaves and stems.

Cholera—A contagious disease.

Chrysalis—*pl.* Chrysalides. The form assumed by some insects before they reach the winged state.

Chrysanthemum—A genus of showy flowering plants belonging to the Compositæ.

Cinchona—A genus of trees yielding quinine.

Circumnutation—The rotating motion made by the growing point of the stem and leaf.

Cissus—A genus of vine-like plants often with brilliant coloured leaves.

Climatic—Influenced by a climate.

Coalesce—To fuse, cohering of parts not usually joined.

Cocos-de-mer—The large double cocoa-nut tree of the Seychelles Isles.

Collomia—A genus of plants whose seeds are remarkable for the spiral fibres which expand elastically when wetted.

Compositæ—A group of plants having an inflorescence of florets arranged upon a common receptacle or head.

Concentric—A number of rings having a common centre.

Cone—The hard woody fruits of the fir-tree.

Coniferous—Fir-like, or cone-like; belonging to the cone-bearing family.

Continuity—Unbroken succession.

Corolla—The second whorl of the floral envelope usually brightly coloured.

Corpuscles—Grains or granular.

Correlation—*i.e.*, connection, interdependence

Cortex—The bark or outer covering of stems.

Cotyledon—A seed leaf.

Cruciferæ—A group of plants having their petals arranged crosswise, with six stamens two of which are longer than the others.

Cryptogamic—Relating to flowerless plants.

Culm—The straw-like stems of the grasses.

Cuscuta—The dodder genus, parasitic upon flax and clovers, &c.

Cuticle—The exterior and thickened part of the epidermis.

Cyclamen—Dwarf primulaceous plants with shortened stems (corms).

D

Dahlia—A genus of tuberous-rooted plants.

Darlingtonia—A genus of Californian plants related to the side-saddle plants.

Datura—The generic name of the thorn-apple.

Débris—Remains, rubbish.

Deciduous—Applied to plants, the leaves of which fall off in autumn.

Dehiscent—Splitting open when ripe.

Deodar—A tree allied to the cedar of Lebanon.

Dentaria—A cruciferous plant bearing bulbils in the axils of the leaves.

Diagrammatic—Drawn to illustrate a statement.

Dicotyledon—A plant whose embryo has two primary seed-leaves.

Diffusion—The intermingling of fluids (gases or liquids).

Diœcious—When the pistillate flowers and staminate flowers are borne upon separate plants of the same species.

Dispersion—Scattering.

Drosera—The generic name of the sundews.

E

Elastic—Springy.

Embryo—The future plant contained in the substance of the seed.

Embryo-sac—The cavity in the substance of the nucellus, containing the egg-cell, which after fertilisation becomes the embryo.

Endocarp—The inside layer of the pericarp.

Entomophilous—Pollinated by insects.

Epicarp—The outside layer of the pericarp.

Epidermis—A layer of generally flattened cells forming the skin of the plant.

Epigean—Developed like the cotyledons of mustard, above ground.

Epipetalous—Growing upon the petals.

Erysipelas—A disease of the blood causing a red eruption.

Eucalyptus—The generic name of the Australian blue gum tree.

Euonomin—A dry extract made from the root-bark of Euonymus atro-purpureus, a North American shrub.

Euonymus—A genus of shrubs and hedgerow trees.

Euphorbia—The spurge genus.

Exogen—Growing by addition to outside of wood and inside of bark, synonymous with dicotyledon.

F

Fermentation—Changes that take place in wort when barm or yeast is added, or when fluids are exposed to the air. *See* Yeast.

Fertilised—Completion of the act of fertilisation, *i.e.*, fusion of the male element contained in the pollen tube with the egg cell of the ovule.

Fibrous—Meaning a structure of fine loose filaments or hairs, *i.e.*, young rootlets.

Fibro-vascular—A compound tissue of fibres and vessels.

Filament—A thread-like fibre.

Flaccid—Want of firmness, soft and lax.

Fructification—The fruit system of a plant.

Fuchsia—A genus of exotic flowering plants having a petaloid calyx.

Function—As applied to plant-life, meaning the use and life-work of the members of a plant.

Fungoid—Growth like a fungus.

G

Gamopetalous—Petals united.

Gamosepalous—Sepals united.

Genesis—Creation, production.

Germinate—The change of the seed from the dormant state to the active growing stage.

Gloxinia—A genus of popular hothouse plants with large handsome flowers.

H

Habitat—The natural abode of a plant.

Herbaceous—Applied to plants which do not form a hard woody stem.

Herbarium—A collection of dried plants.

Hexagonal—A six-sided and angled figure.

Hibernating—Sleeping, a dormant condition.

Hilum—The black scar on a bean seed.

Hippuris—A genus of aquatic flowering plants.

Horizontal—Parallel to the horizon level.

Hoya—A genus of tropical climbing plants.

Hyacinths—Bulbous plants.

Hydrangea—A genus of flowering shrubs.

Hygienic—Relating to the preservation of health.

Hygrometric—Moisture and its influence.

Hygroscopic—Applied to the film of water surrounding the particles of the soil.

Hypericum—The generic name of the St. John's wort.

Hyphæ—Filaments or threads of the fungus spawn.

Hypogean—Development of the cotyledons under ground.

I

Impatiens—The generic name of the balsam.

Impervious—Not to be penetrated by water.

Insectivorous—Catching and killing insects, plants that have this power and can absorb the decomposed insects.

Insoluble—Substances that do not dissolve in water.

Intercellular—Spaces between the cells.

Internode—The space between two nodes.

Involucre—A whorl of bracts.

Iodine—A soluble substance extracted from kelp and used as a test for starch.

L

Laburnum—Yellow-flowered trees allied to the Pea family (Leguminosae).

Legume—The dehiscent fruit of the pea family, a pod.

Leguminosae—A family of plants having for their fruit a legume or pod, i.e., Pea, Laburnum.

Lenticels—Minute pores in the bark.

Liane—A hanging root or stem.

Liber—The inner bark, same as phloëm.

Linnaea—A genus of dwarf trailing plants.

Luscious—Sweet and succulent.

M

Magnesium—The metallic base of magnesia.

Magnolia—A genus of flowering shrubs and trees.

Mahonia—A genus of evergreen shrubs belonging to the barberry family.

Martynia—A genus of plants having capsules with long curved hooks.

Melampyrum—A genus of dwarf flowering plants partly parasitic.

Membranous—Thin and destitute of green colour usually applied to bracts.

Mesocarp—The central layer of the pericarp.

Mesophyll—The ground tissue of the leaf.

Metabolism—Changes which take place in protoplasm and which it causes in other substances.

Microbes—A term applied to one-celled plant atoms, like bacteria.

Micropyle—A small pore in the coats of the ovule through which the pollen tube passes.

Modicum—Moderate sized ; a small quantity.

Monocotyledon—A plant whose seed is furnished with one seed leaf.

Monœcious—Applied to a plant when the stamens and pistil are in distinct flowers.

Monstera—A genus of climbing aroids with edible fruit.

Mucilaginous—Sticky, gumlike, secreting mucilage.

Mucuna—A genus of Brazilian Leguminosæ, yielding the cowage (consisting of intensely irritating hairs), of the Materia Medica.

Mutualism—Interchange of some advantage, botanically applied to the union of two dissimilar plants which live in *close* contact with each other to their mutual benefit.

Mycelium—The root-like colourless filaments of fungi.

Mycetozoa—A term applied to the slime-fungi.

N

Nectary—A honey secreting gland or spur.

Nemophila—A genus of dwarf annual flowering plants.

Nepenthes—A genus of plants having as a prolongation of the midrib of the leaves, ascidia or pitchers.

Nocturnal—Happening by night.

Node—The exact point on the stem from which the leaf is developed.

Normal—Regular, unaffected by any modification.

Noxious—Hurtful or poisonous.

Nucellus—The internal tissue of the ovule within which the embryo-sac is embedded.

Nutrition—The process and function of taking in food for the purpose of growth and to replace waste.

O

Orchis—A genus of the orchid family growing in the soil.

Osmosis—The passage of fluids through a membrane.

Ovary—The ovule case, that part of the carpel that bears ovules.

Ovule—The structure which after fertilisation forms the seed.

Ovum—The egg cell of the ovule.

Oxalis—The generic name of the wood sorrel.

Oxygen—A gas, one of the constituents of the atmosphere.

P

Palisade-tissue—A tissue of oblong cells placed side by side at right angles to the flat surface of the leaf.

Papilionaceous—Butterfly shaped.

Pappus—A light hairy development from the calyx of some plants.

Parasite—The habit of growing upon and deriving nourishment from another plant.

Pellucid—Shining and transparent.

Perennial—Plants that live for an indefinite period.

Perianth—A term used when there is no distinction between calyx and corolla.

Pericarp—The ripened walls of the ovary constituting the structure of the fruit.

Persistent—Applied to the parts of the flower that remain on for some time.

Petunia—A genus of Brazilian Solanaceae.

Philodendron—A genus of aroids usually climbers.

Phleum—A grass.

Phloëm—The inner bark, containing sieve-tubes.

Phosphate—A salt formed by the union of phosphoric acid with some base.

Phyllotaxis—The law of leaf arrangement.

Physalis—The generic name of the winter cherry.

Physiological—Having reference to the function or life work of the plant.

Picea—A genus of the Conifer family.

Pinetum—A garden devoted to the culture of pine-trees.

Pinguicula—The generic name of the butterworts.

Pinus—A genus of the Conifer family.

Pistil—The female part of the flower consisting of ovary, style, and stigma.

Pistillate—Applied to flowers having the pistil only.

Pith—The soft tissue in the centre of the stem.

Plumbago—The generic name of the leadworts, small flowering plants and shrubs

Plumule—The first stem shoot of the germinating seed.

Poa—A grass.

Poinsettia—A genus of Mexican plants having bright scarlet bracts and small flowers.

Pollard—A tree trunk with its branches cut short.

Pollen—The fertilising or male part of the flower.

Pollination—The act of conveying the pollen from the stamen to the stigma.

Polypetalous—Separate or many petals.

Polysepalous—Separate or many sepals.

Polytrichum—The generic name of the hair moss.

Potassium—The metallic base of potash.

Proboscis—The feeling and feeding organ of an insect.

Prothallus—The first growth when the spore of a fern germinates.

Protococcus—A genus of unicellular plants forming a green stain upon trees, &c.

Protonema—The first growth of the moss-spore.

Protoplasm—A highly complex substance forming the essential part of all living cells, and to which all life growth is due.

Prototypes—First forms of plant-life.

Psamma—A genus of the grass family.

Pseudo-bulb—A swollen **stem** common in the epiphytic orchids.

Pteris—The generic **name of the** bracken fern.

Q

Quiescent—Inactive, **dormant.**

Quinine—An **alkaloid** extracted from **the cinchona trees.**

R

Radicle—**The first formed root when a seed germinates.**

Rafflesia—A **genus of brown leafless parasites.**

Receptacle—**That part of the stalk on which the flower is developed.**

Resin—A **secretion from certain trees which hardens on** exposure.

Respiration—**The process of breathing.**

Rhananthus—**The generic name of the yellow-rattle (a root parasite).**

Rhododendron—A genus of popular flowering shrubs and **dwarf trees.**

Root-cap—A loose covering of tissue that protects the extreme **point of the growing root.**

Root-hairs—**The delicate unicellular hairs found on the young root.**

S

Salicine—A substance obtained from the bark of willows, **soluble in water and alcohol, and crystallising in** bright **white needles.**

Salvia—A genus **of labiate plants.**

Samara—Winged **fruit.**

Saprophyte—Plants **that live upon decaying organic matter.**

Sarracenia—**The generic name of the North American side-saddle plants.**

Saxifraga—A **genus of dwarf** Alpine plants.

Scales—Rudimentary leaves.

Secretion—Applied to substances like resin and honey, the production of assimilation and metabolism.

Sedum—A genus of succulent Alpine plants.

Soluble—Any substance that dissolves in water.

Spadix—The inflorescence of the Aroideæ.

Spathe—The bract of the Aroideæ.

Sporangium—The spore-case of some of the cryptogamia.

Spurious—False.

Stapelia—A genus of succulent plants, very poisonous and fœtid.

Starch—Colourless grains, a product of assimilation in the leaf.

Stigma—The receptive part of the pistil.

Stipa—A genus of the grass family.

Stipules—Small outgrowths at the base of the petiole.

Stomata—Minute pores in the epidermis of the leaf or green stem.

Sulphate—A salt formed by the combination of sulphuric acid with some base.

Sycamore—The plane tree of Scotland, Acer pseudoplatanus.

Symbiosis—Mutualism, a living for one another, interchange of benefits by united growth.

Syncarpous—United carpels.

T

Tannin—A substance widely diffused through the leaves and stems of plants, of an astringent character.

Tap-root—A root that forms an unbranched tapering axis : Ex., carrot.

Tendril—A coiled or hooked filament modified to assist plants to climb.

Tentacles—The glandular and feeler-like hairs of the sundew.

Terminal—At the apex or end.

Testa—The skin of seed.

Tigellum—The first stalk of the seed bearing the cotyledons.

Tillandsia—A New World genus of perching or epiphytic plants.

Tissue—A group of cells; having a common origin.

Tormentilla—A genus of small creeping rosaceous plants.

Transpiration—The giving off of water vapour from the surface of leaves and stems.

Tricyrtis—The generic name of the toad-lily.

Tuber—A fleshy root or succulent underground stem.

U

Umbelliferæ—A group of plants having an umbellate arrangement of the inflorescence or flower-head.

V

Vallisneria—A genus of aquatic flowering plants.

Valved—Having valves, e.g., anther of the barberry.

Vapour—Gas into which most liquids and solids are converted by heat.

Vasculum—A little vessel or box for collecting botanical and other specimens.

Venation—The arrangement of veins in a leaf.

W

Weigelia—A genus of flowering shrubs allied to Honeysuckle.

Whorl—An arrangement of leaves or parts of the flower in rings.

Wort—Sweet unfermented new beer.

Y

Yeast—A unicellular plant that sets up fermentation under certain conditions.

www.ingramcontent.com/pod-product-compliance
Lightning Source LLC
Chambersburg PA
CBHW021404210326

41599CB00011B/1008